SMARTEST PREP

Guide to LSAT Logic Games

Pratheep Sevanthinathan
and
Padmini Raghavan

Illustrated by Shantel Miller

University Press of America,® Inc.
Lanham · Boulder · New York · Toronto · Plymouth, UK

4501 Forbes Boulevard
Suite 200
Lanham, Maryland 20706
UPA Acquisitions Department (301) 459-3366

10 Thornbury Road
Plymouth PL6 7PP
United Kingdom

Printed in the United States of America
British Library Cataloging in Publication Information Available

Library of Congress Control Number: 2013952419
ISBN: 978-0-7618-6271-0 (paperback : alk. paper)
eISBN: 978-0-7618-6272-7

™The paper used in this publication meets the minimum
requirements of American National Standard for Information
Sciences—Permanence of Paper for Printed Library Materials,
ANSI Z39.48-1992

Dedication

This book is dedicated to my family – Andrea, Sevanna, Mom and Appa – for your support and love throughout the years. This project would never have been possible without my wife's absorption of our day-to-day stresses, and she deserves special recognition for this. Thank you so very much Andrea.

I would like to give an extra special thank you to my father-in-law, Subhash Sethi, for giving me the confidence and peace of mind to focus on this project. You will be missed.

- *Pratheep Sevanthinathan*

This book is dedicated to my family –Mom, Dad, and Preethi. Thank you for allaying my stress, tolerating my mess, and rooting for my success. I could not have done this without you.

- *Padmini Raghavan*

Acknowledgements

We would like to thank Shantel Miller for her illustrations throughout this manual, and to Alexis Sharp for her tireless proofreading, feedback, comments, and assistance. This book would not have been possible without your help.

CONTENTS

Section I

GETTING STARTED

CHAPTER 1 – WHY smarTEST PREP?

In this Chapter you will learn:

- Why smarTEST Prep is different.
- How smarTEST Prep can benefit you.

LESS IS MORE

The Analytical Reasoning Section of the LSAT, commonly called the Logic Games Section, can be an immense hurdle for anyone seeking admission to law school. This single section has the ability to control your performance on the entire LSAT and, ultimately, your ability to get into the law school of your choosing. For this reason, there are a plethora of guides, courses, and manuals specifically dedicated to the Logic Games Section. With the wealth of choices you may wonder, "Why smarTEST Prep?"

The answer to that question: *simplicity*. In a sea of increasingly complex test preparation, we offer a river of clarity. smarTEST Prep bucks the trend for test prep manuals that require you to learn overly complicated or proprietary methods to answer questions. 5-step, 7-step, or 10-step methods can be distracting, confusing, and ultimately counterproductive. What happens if you miss a step when applying such a process? What happens if you cannot remember every step on test day? You could potentially blow an entire game and cost yourself a shot at your dream school.

Test prep manuals should not add to your stress on test day. To the contrary, your test prep manual should *simplify* your studies. Analytical Reasoning on the LSAT, as tested by logic games, derives from an important, intuitive, and foundational concept – logic. The smarTEST Prep Guide to LSAT Logic Games aims to simplify and streamline your studies by training you to uncover the hidden but accessible logic behind each game.

ONE SIZE DOES NOT FIT ALL

Students preparing for the LSAT do not all think alike, work alike, or perform alike. When using one-size-fits-all test prep courses, some students can feel left behind, while others can be stifled by a slow pace. This guide aims to provide targeted strategies and tools to help you earn a better Analytical Reasoning score regardless of where you currently stand.

Specifically, students who are struggling will benefit from Section II: Logic Game Basics, which thoroughly breaks down the elements of logic games and the important task of diagramming. Students who understand the concepts but need to improve their technique can move on to Section III: Strategy. Advanced students will benefit from the some of the more complex games provided in Section IV of this book. For more details see Chapter 2: How to Use This Book.

UNDERSTAND THE ELEMENTS

This manual, unlike most others, focuses on the foundation and fundamentals of a logic game. In Section II: Logic Game Basics, we break down the elements of a logic game and give you step by step explanations on how to solve a game. We do not make assumptions about your knowledge. We want to help you dissect logic games to determine exactly what the LSAT writers were trying to accomplish when creating a game. You must know what you are reading, why you are reading it, and what you should be looking for when approaching a logic game.

This book will help you gain a strong understanding of the foundation, components, and purpose of a logic game by teaching you every aspect of how a logic game is comprised and written. You will see the structure and patterns that are common among the games, and recognize the nuances of the Logic Games section.

PREPARE FOR THE UNEXPECTED

Encountering unfamiliar game types or questions is common on the LSAT, and can wreck your momentum and confidence if you encounter them on test day. Further, common logic game types and questions may appear unfamiliar under the stress and time constraints of test day. The smarTEST Prep Guide to LSAT Logic Games has set out to prepare you for these exact types of games and situations.

We present a methodical approach to attacking and solving unfamiliar logic games (See Chapter IV: The Game Directive). Although we ensure that you become familiar with common game types, we also ensure that you do not become overly dependent on methods that apply solely to these common game types. This book is geared to prepare you to solve any logic game, whether or not you recognize the game type.

ORIGINAL GAMES

Many test prep materials include actual old LSAT questions licensed from the makers of the test. While working with actual LSAT questions in a simulated practice exam setting is recommended, it is not advisable to use these questions when studying. If you study with actual LSAT questions there is a strong possibility that you will encounter these questions again when taking practice tests. This, in turn, can taint your LSAT practice exams and give you artificially inflated scores. Ultimately, using actual LSAT questions during studies *and* simulated practice exams could lead to an inaccurate understanding of your abilities.

Accordingly, Section IV of this manual includes several dozen original LSAT logic games that are representative of actual LSAT logic games. Our logic games have never been published anywhere else and provide test takers with all new questions. We have gone through great lengths to ensure that the games are not only representative of the actual LSAT, but that they vary in difficulty, length, and type. Further, our solutions are detailed and provide full, thorough, and step-by-step explanations on how to answer every question presented. Our original logic games can be used to focus on weaknesses and streamline your studies to improve your technique. You can then implement what you have learned through our original games when taking practice exams with actual LSAT questions.

COMPLEX LOGIC GAMES

Several of the logic games in Part 2 of Section IV are challenging and more complex than what would typically appear on the LSAT. Working through these difficult games will be highly beneficial in your training. Practicing with complex games is analogous to "swimming with a drag suit."

Competitive swimmers often wear cumbersome baggy swimsuits over their regular swimsuits to create extra resistance and drag in the water during practice. On competition day, without the drag suit, the swimmers apply the same effort and technique they learned when practicing with the drag suit, but will glide through the water with ease, performing much better since there is no extra resistance and drag. Likewise, practicing with complex LSAT logic games questions will be a difficult but educational task. If you can handle these difficult games in practice, you will be well prepared to handle the games you encounter on test day.

CHAPTER 2 –HOW TO USE THIS BOOK

In this Chapter you will learn:

- How to best utilize this book, given your current scoring range.

DETERMINE WHERE YOU STAND

It is important that while studying for the LSAT, you are cognizant of your current scoring range, areas of strength, and areas in need of improvement. The smarTEST Prep Guide to LSAT Logic Games will help you determine and address these factors to put you in the best position to succeed on the LSAT.

With that being said, there is no substitute for doing actual practice LSAT exams, under actual LSAT conditions. We strongly recommend that you periodically take full practice LSAT exams, under exam conditions (i.e., timed, in a quiet room, no distractions, and no pausing for breaks). Actual LSAT exams are released by the Law School Admission Counsel and can be purchased directly from the LSAC or from online book retailers – look for the 10 Actual, Official LSAT Prep Tests by the LSAC.

In the final two months leading up to your scheduled LSAT, you should try to take a full practice LSAT every week. This will not only give you an accurate understanding of your strengths and weaknesses, but will also get you in a routine that can be applied on test day. Having a routine is essential to calm your nerves and maintain composure on test day.

STREAMLINE YOUR STUDIES

It is fully advised that anyone planning on taking the LSAT read this *entire* book *thoroughly*. However, we completely understand that students are diverse in their abilities, progress, and concerns. The guidelines below are a cheat sheet for using this manual, based on your current LSAT score. Once you have determined where you stand, use these guidelines to streamline your studies and use of this manual.

The ranges provided are subjective guidelines, and may not apply *exactly* to you. Feel free to follow the guidelines for the ranges slightly above or below your score.

If you are scoring between 120-149; or
If you struggle to complete your Games sections:

We recommend that you strengthen your fundamentals and then build speed and accuracy.

- Read through the entire book.
- Pay special attention to Section II: Logic Game Basics.
    - Understand *why* each element of diagramming is important.
    - Pay special attention to Chapter 4: The Game Directive
- Take a past Logic Games section from a real LSAT and work through a game using the analysis in Section II. Identify the background, rules, and question types. Isolate the Game Directive, and make deductions. Work through the elements of your diagram alongside the example in the book.
- Read through Appendix A: Formal Logic.
- Read through Appendix B: Symbology

- Work through Section IV: Games Part 1 at 7 minutes per game.
- Return to Section III and review again before moving on to Games Part 2.
- Work through Section IV: Games Part 2 at 10 minutes per game.
- Remember to thoroughly review the solutions to the logic games you complete!

If you are scoring between 150-159; or
You score 12-18 on your Game sections:

You probably understand the basics, but can unlock your scoring potential through smart strategy.

- Read the entire book.
- Pay special attention to Chapter 4: Game Directive. Make sure you know how to deal with unexpected game types.
- Pay special attention to Section III: Strategy. This entire section is the key to developing a streamlined strategy to save time and efficiently earn points.
- Take a past Logic Games section from a real LSAT and review your strategies in approaching the Games. Compare your approach with the one offered by the book. Look for ways to simplify your approach to games.
- Review Appendix A on Formal Logic.
- Work through Section IV: Games Part 1 at 6 minutes per game.
- Work through Section IV: Games Part 2 at 9 minutes per game.

If you are scoring between 160-165; or
You score 15-20 on your Game sections:

You are likely doing well on familiar and recognizable games, but may be losing points on hybrid or unexpected games. We recommend you increase speed on your best games, and increase accuracy on your weaker games.

- Skim Section II to learn symbols or methods to increase speed.
    - Read Chapter 4: The Game Directive. Learn to deal with unexpected game types.
- Skim Section III: Strategy. Confirm that you are using your best strategies for working through the games.
- Read Chapter 9. Confirm that you are using key tips for specific game types.
- Read Chapter 10 carefully. Incorporate strategies to streamline your approach.
- Work through Section IV: Games Part 1 at 5 minutes per game.
- Work through Section IV: Games Part 2 at 8 minutes per game.

If you are scoring between 166-172; or
You score 20 or more on your Game sections:

You are in good shape but must keep your momentum and perform consistently to ensure you will deliver on Test Day.

- Read Chapter 4: The Game Directive.
- Skim Section III.
- Skim Chapter 9.
- Read Chapter 10.
- Work through Section IV: Games Part 1 at 4.5 minutes per game.
- Work through Section IV: Games Part 2 at 7 minutes per game.

If are scoring above 172; or
You score 22-24 on your Game Sections:

Prepare for the unexpected and the unfamiliar to reach perfection on test day.

- Skim Chapter 4.
- Read Chapter 10.
- Work through Section IV: Games Part 1 at 4 minutes per game.
- Work through Section IV: Games Part 2 at 6 minutes per game.

Section II

LOGIC GAME BASICS

CHAPTER 3 – ANATOMY OF A LOGIC GAME

Familiarize yourself with the components of a logic game: Background, Rules, and Questions.

CHAPTER 4 – THE GAME DIRECTIVE

Learn how to use Game Types to speed up, and use the Game Directive to deal with unexpected Games.

CHAPTER 5 – DIAGRAMMING

Learn the elements and objectives of diagramming.

CHAPTER 6 – DEDUCTIONS

Learn how to be methodical in making key deductions.

CHAPTER 7 – TESTING ANSWER CHOICES

Learn how to test answer choices to find the correct answer.

CHAPTER 3 – ANATOMY OF A LOGIC GAME

In this Chapter you will learn to:

- Understand the main components of a logic game: background, rules, and questions.
- Identify common rule types.
- Learn common question types.
- Approach the questions in the most efficient order.

WHAT ARE LOGIC GAMES?

Logic games are complex puzzles of logical statements that comprise the Analytical Reasoning section of the Law School Admission Test ("LSAT").

- 35 minutes
- 4 games
- 22-24 questions

Test takers are allowed 35 minutes to work through four games and answer 22-24 questions. The complexity of these puzzles, combined with the time crunch, can be daunting on test day and will often influence a student's performance throughout the entire LSAT. It is no surprise that the Logic Games section of the LSAT can be the most challenging to many law school applicants.

Yet there is no need to fear logic games – there is a method to the madness, and conquering logic games is a realistic possibility for all LSAT takers.

This book will guide you through the steps required to:

- Analyze a logic game
- Diagram the rules
- Work through questions efficiently
- Lock in the correct answers

By following the guidelines set forth in this book, and through disciplined practice, you should be able to confidently work through *any* logic game presented on test day.

ENTITIES, SCENARIOS, AND RULES

At its core, an LSAT logic game is a complex puzzle that requires you to negotiate *entities* within a *scenario* according to certain *rules*. For example, a game may ask you to place a student's classes (the entities) into a particular schedule (the scenario) subject to certain constraints and restrictions (the rules). You are given a set of rules and have to abide by them in order to accomplish your goal of answering the questions. When scheduling the classes, one rule may be that the student always has to attend math class before history class. Hence, as you figure out the order of the schedule you must always remember math comes before history. That is where the "game" aspect comes into play – there are limitations to how the puzzle can be solved, and it is your job to navigate within these limitations as you work through the scenario.

LSAT logic games are much like other familiar games that are defined by logic and rules. Below are some examples of popular household games, which contain logic similar to what we see in the LSAT Logic Games section.

- In Chinese Checkers, pieces (entities) can move one space at a time into empty adjacent spaces (rule), or may jump over other pieces consecutively (rule) to move across the board towards the goal (the scenario).

- In Sudoku, a partially filled grid (scenario) must be filled with digits (entities) so that each column, each row, and each subgrid contains all the digits from 1 to 9 (rules).

- In Chess, a King (entity) can move one space at a time in any direction (rule), but a Knight (entity) can only move in an "L" shape (rule) on the chessboard (scenario).

Similarly, on an LSAT logic game, entities are subject to rules that limit their abilities. In the previous example, history class was limited by its requirement to follow math class. History class is unable to precede math class, and can also never be the first class in the schedule. It is your job to abide by these rules when navigating the logic game.

Logic games can vary in complexity, length, structure, and focus. However, they are all presented in very similar manners and tend to share key common elements that allow for methodical processing. There are three main parts to every LSAT Logic game: the *background*, the *rules*, and the *questions*. These parts will be examined in detail below.

THE BACKGROUND

The *background* of a logic game is contained in the opening paragraph and will describe the *entities* and *scenario* that are the focus of the game.

THE ENTITIES

Entities are the subjects of a logic game – the focus of every scenario and question. Entities are usually people or objects, but can be just about anything (e.g., songs, languages, years, cities, and recipes). Typically, the entities are provided in a list, separated by commas, and often appear between two dashes.

Every logic game has at least one set of entities, but multiple sets are not uncommon. Customarily, the entities will be presented in alphabetical order and when there are two sets, the sets will often be comprised of entities starting with letters at opposite ends of the alphabet.

Example 3.1 - Entities

> Seven soccer players on a team are selected to participate in a shootout. Three players are forwards – Adam, Bill, and Carl – and four players are defenders – Quincy, Rick, Sam, and Tim.

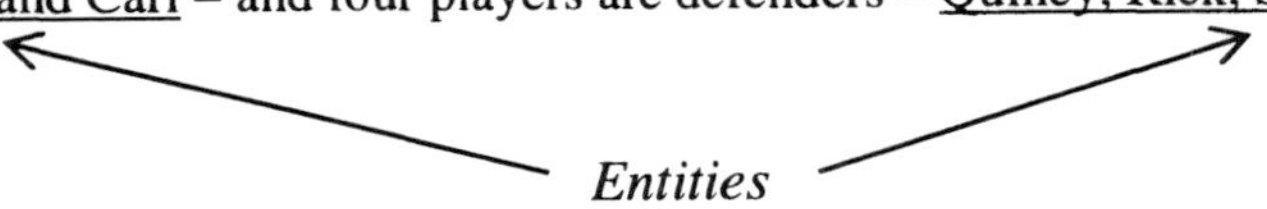

In the above example, the forwards comprise one set of entities, and the defenders comprise a second set of entities. Notice how *Adam, Bill, and Carl* are separated from *Quincy, Rick, Sam, and Tim.* This should immediately tip you off that *A, B, C,* as a group, behave differently than *Q, R, S,* and *T.* The members of each set share a common attribute that differentiates them from the other set.

THE SCENARIO

The scenario of the game is what the entities are doing or what is being done to the entities, i.e. the ultimate goal for the entities. This most often entails placing the entities in an order, placing entities in spots, selecting certain entities from a group, matching entities, grouping entities, and separating entities. The scenario begins broadly, with an abundance of ways to place entities, and is then limited by subsequent rules.

Example 3.2 – A Scenario

> Seven soccer players on a team – Adam, Bill, Carl, Dennis, Ethan, Fred, and Greg – are selected to participate in a shootout. Each player will shoot exactly once and will shoot alone. The order in which the players shoot must be in accordance with the following conditions:

THE RULES

The rules of a logic game follow immediately after the background, and are a set of specific statements that define and limit the scenario. The rules state what the entities must, can, and cannot do. There are typically two to seven rules for each logic game, which must be viewed as an aggregate set that applies throughout the game. Each rule applies to every question and no rule can be ignored. When viewed jointly, the rules provide key deductions to further limit the scenario.

Example 3.3 – A Set of Rules

> Seven soccer players on a team – Adam, Bill, Carl, Dennis, Ethan, Fred, and Greg – are selected to participate in a shootout. Each player will shoot exactly once and will shoot alone. The order in which the players shoot must be in accordance with the following conditions:

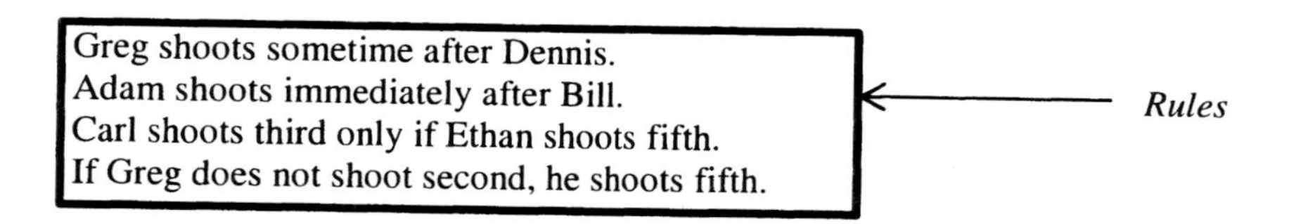

POSITIONAL RULES: RELATIVE TO A SPOT

Rules will focus on an entity, or multiple entities, relative to spots or other entities. When a rule is written relative to a spot, it is a *positional rule*. The last two rules above – "Carl shoots third only if Ethan shoots fifth," and "If Greg does not shoot second, he shoots fifth," – are positional rules because they mention spots and allow for the placement of entities into spots.

RELATIONAL RULES: RELATIVE TO OTHER ENTITIES

Rules that are not written relative to a spot will relate entities to one another. These rules are *relational rules*. The first two rules above – "Greg shoots sometime after Dennis," and "Adam shoots immediately after Bill," – are relational rules because no spots are mentioned. Distinguishing between positional and relational rules will be important when we start diagramming later in this book.

The ability to comprehend, manipulate, and apply rules is the essence of what the LSAT is trying to test in the Games section. When reading the rules, contemplate their importance and influence on the scenario. Think about which entities are involved and what exactly the rule is telling us to do with the entities. After gaining a full understanding of the rules, deductions can be made and solving the logic games will become much easier.

READ THE RULES CAREFULLY

Always keep in mind that to score high, you must cautiously and critically ready every word in the rule. Simple mistakes such as skipping a rule or glancing over the words "only" or "not" could easily *cost you the entire game* and six important LSAT points. Six points could mean the difference between your dream school and your safety school!

THE QUESTIONS

The questions of a logic game test your understanding of the rules and their interplay with the scenario. Most significantly, you will be tested on your comprehension of the logic behind the rules and your ability to make key deductions.

QUESTIONS MAY ADDRESS KEY DEDUCTIONS.

For instance, based on the game in Example 1.3, a deduction should have been made that Dennis can never shoot fifth, sixth or seventh (the deduction will be explained later in this section). A question that focuses on this deduction may ask:

Example 3.4 – Question

Each of the following could be true, EXCEPT:
(A) Dennis shoots first
(B) Dennis shoots third
(C) Adam shoots seventh
(D) Dennis shoots sixth
(E) Adam shoots fifth

While you could answer this question through trial-and-error, you would waste valuable time. However, once you make the deduction that Dennis cannot shoot sixth prior to answering the questions, the correct answer, (D), is obvious. This is the heart of what the LSAT Games section is testing.

QUESTIONS MAY TEST YOUR UNDERSTANDING OF THE BOUNDARIES OF THE RULES.

The questions will also test your ability to identify the boundaries of the rules and whether you are viewing the rules as a whole, without omitting any key information. Such questions would ask what are the most, least, and exact number of possibilities that the entities can be arranged within the specified rules. (E.g., *How many* different shooting orders are possible if Carl shoots third?)

QUESTIONS MAY POSE ADDITIONAL LIMITATIONS OR CHANGE THE RULES.

Lastly, some questions will add additional limitations and some may even ask you to temporarily change a rule. These types of questions are introduced to test your ability to adapt and quickly apply new logic. (E.g., If the rule that George shoots sometime after Dennis is removed, and replaced with the rule that Fred shoots exactly one spot before Dennis, which of the following could be true?)

As indicated above, questions can vary greatly in form and substance. However, the questions can be generally classified into six main types. Understanding and becoming familiar with these common question types can dramatically improve your score by giving you a head start on each question. Each question type has specific wrong answers that can be easily identified and eliminated. Going into the LSAT with a strong grasp of common wrong answers is sure to save you time, and will help narrow down answer choices immediately.

The following table describes common question types, along with indicators of right and wrong answers:

Table 3.1 – Common Question Types

1. Complete and accurate (acceptable)
   Example: Which of the following is an acceptable order of entities?
   Right Answer: no rules are violated
   Wrong Answers: at least one rule will be violated

2. Could be true (can be true, must be false except)
   Example: Which one of the following could be true?
   Right Answer: could be true
   Wrong Answers: must be false

3. Could be false (can be false, must be true except)
   Example: All of the following must be true EXCEPT
   Right Answer: could be false
   Wrong Answers: must be true

4. Must be false (cannot be true, could be true except, can be true except)
   Example: Each of the following could be true EXCEPT
   Right Answer: must be false
   Wrong Answers: could be true, must be true

5. Must be true
   Example: Which one of the following must be true?
   Right Answer: must be true
   Wrong Answers: could be false or must be false

6. Numbers (minimum, maximum, least, greatest)
   Example: What is the maximum number of classes that can be attended by the students?
   Right Answer: requires calculation/deduction
   Wrong Answers: requires calculation/deduction

The above question types have been ordered from the easiest to hardest, typically, based on how quickly one should be able to come to an answer. However, the actual difficulty of a question involves more than just the type of question. Other factors include: how many deductions are required to answer the question, which rules are implicated by the question, which entities are implicated by the question, and whether a new diagram is required.

Some questions you will encounter can be answered with little to no work, while others will require you to try five separate configurations. How much work you will have to do for each question will depend on a multitude of factors including:

- The type and difficulty of each question;
- The type and difficulty of each game; and
- Whether you have already worked through the configuration(s) in previous questions

The question type is the most influential of these factors. "Must be false" and "Could be true, EXCEPT" questions will often require you try the situations presented in all five answer choices. These are time consuming questions that should, generally, be answered last. Conversely, "Complete and accurate list" (acceptability) questions can usually be answered solely with your diagram, without having to work through various configurations or writing anything down.

CHAPTER 4 – THE GAME DIRECTIVE

In this Chapter you will learn to:

- Use common game types as a tool not a crutch.
- Use the Game Directive to deal with unexpected games.

GAME TYPES V. GAME DIRECTIVES

One of the great challenges in the Games section is the variety of the presentation. Very simple concepts are often tested in complicated ways through an assortment of logic game types. Each one of these "types" requires a different thought process and different diagrams. The variation in the game types can therefore pose a significant challenge, especially when the type is not easily identifiable.

You can save valuable time on test day by identifying game types and using the strategies and diagrams associated with those types. Accordingly, familiarizing yourself with the types of games commonly presented on the LSAT can be highly beneficial. This is precisely why the foundation of most test prep courses and materials on logic games centers on the classification and differentiation of game types.

However, what if you are unable to identify the game type? What if you come across an unrecognizable game type or, under the stress of test day, you are unable to recognize a familiar game type? Such a situation is common, and will often cause anxiety and confusion that could potentially linger throughout your entire test.

This manual will equip you to avoid this troublesome situation. If you cannot identify the game type, you must still take action on the game. The solution: instead of focusing on the type of game, we urge you to understand the logic entrenched in the game by focusing on the *Game Directive*.

GAME DIRECTIVE: ORDER ENTITIES, MATCH ENTITIES, OR POSITION ENTITIES.

Betting that you will recognize the game type for each game you encounter can be a recipe for disaster. To avoid this, focus on what the game is asking you to do. Every game contains a directive – what the game is asking you to do – which you should look to identify *before* you attempt to determine the game type. Virtually all games will essentially ask you to do one or more of only three directives: (1) order entities, (2) match entities, or (3) position entities. Once you determine which of these directives you are tasked with, you will be well on your way to finding the correct answers.

FAMILIARIZE YOURSELF WITH GAME TYPES AS A LEARNING TOOL, NOT A CRUTCH!

While we strongly urge you to focus on the Game Directive, you should still be familiar with the common game types that exist on the LSAT. If you are able to recognize the game type and are familiar with the corresponding diagram and special deductions, you will save valuable time on that game. The table below provides an overview of the most commonly occurring game types. Spend some time reviewing this table and familiarizing yourself with these game types with the understanding that this is not a conclusive listing of game types.

Table 4.1 – Common Game Types

Type	Key Features	Example	Frame
Assigned Ordering: **(Game Directive: Order Entities)**	Entities are ordered according to defined spots.	Seven basketball teams – Jaguars, Kings, Lions, Mavericks, Nighthawks, Owls, and Pistons – will be seeded 1 through 7 for a tournament. The seeding must follow the following conditions: The Lions are seeded third.	1 2 3 4 5 6 7
Relative Ordering: **(Game Directive: Order Entities)**	Entities are ordered relative to other entities.	Seven basketball teams – Jaguars, Kings, Lions, Mavericks, Nighthawks, Owls, and Pistons – will be seeded for a tournament. The seeding must follow the following conditions:	C…F A…B D…E…G 1 2 3 4 5 6 7
Matching: **(Game Directive: Match Entities)**	Entities are matched with spots	Seven players – Adam, Ben, Chris, David, Evan, Frank, and George – play for the Jaguars. Two players are guards, three players are forwards, and two players are centers. The players will be assigned positions according to the following conditions:	Guards Forwards Centers
Selection & Grouping: **(Game Directive: Match Entities)**	Entities selected from a group of entities	Seven teams – Jaguars, Kings, Lions, Mavericks, Nighthawks, Owls, and Pistons – are competing for four spots in a basketball tournament. The four spots will be chosen according to the following conditions:	No Frame
Geometric Positioning: **(Game Directive: Position Entities)**	Entities are placed around a shape	Six people – Adam, Ben, Fred, Evan, David, and George – are sitting equally spaced around a circular table. The seating assignments follow these conditions:	A B F E D G
Mapping: **(Game Directive: Position Entities)**	Entities are placed on a map based on coordinates or specific locations	Gotham City is split into four evenly sized boroughs – Brooklyn is in the northeast, Scarborough is in the northwest, Hollywood is in the southeast, and the French Quarter is in the southwest. A franchise plans to build nine coffee shops in Gotham City according to the following conditions:	Scarborough Brooklyn French Hollywood

Again, Table 4.1 shows only the common game types that appear on the LSAT. *The table is not comprehensive and does not show every game type you will encounter.* In fact, there is a very good chance that you will come across a game that is a combination of one or more of the above game types. That is precisely why this book focuses on directives and not types. More details about how to determine the Game Directive, diagram, and make special deductions are covered in Section III.

UNKOWN GAME TYPES

One of the biggest mental jabs that test-takers can suffer on test day is finding a logic game that they have "never seen before." Far too many test-takers end up blowing an entire game on test day because they cannot figure out the "type" of game they are dealing with. This is can be crippling to your LSAT composite score, *as misfiring on an entire game will cost approximately six points on the LSAT.* Six points could mean the difference between your dream school and your safety school!

One explanation for this common occurrence is that many other courses will overstate the importance of training your mind to immediately pick out the "type" of game. True, it is important to know about the different types of games and how the logic works with each type. However, do not fall into the trap of believing every logic game presented will fall nicely into a defined game type. They certainly will not, and you do not want to discover this for the first time on test day.

In reality, it is more important to be able to identify the logic entrenched in each game and the interplay between entities and scenarios (the directive). Knowing this will allow you to deal with and diagram the logical scenarios and rules no matter how they are presented. *Understand common game types to save time. Use the Game Directive to take action on unfamiliar games.* By adhering to this basic tenant you will put yourself in the best position to attack any and all games that are presented.

CHAPTER 5 - DIAGRAMMING

In this Chapter you will learn:

- ❖ The elements of diagramming.
- ❖ The objectives of diagramming.

WHAT IS DIAGRAMMING?

Diagramming is the single most valuable technique used to solve logic games. It is the act of transforming the words that comprise the scenario and rules into shorthand notations and figures. Good diagrams will be comprehensive, yet streamlined. They will reduce an entire logic game into illustrated instructions, a-la Ikea furniture manuals. Ultimately, *everything* you know about a particular logic game should be condensed into its corresponding diagram so that the answers to the LSAT questions will be staring right at you on test day.

The following example shows how a matching game can be reduced into an illustrated guide. Do not worry if you are unable to interpret this diagram or figure out how it was conceived. The chapters that follow will cover, in detail, how to make and use such diagrams.

Example 5.1 – Completed Diagram

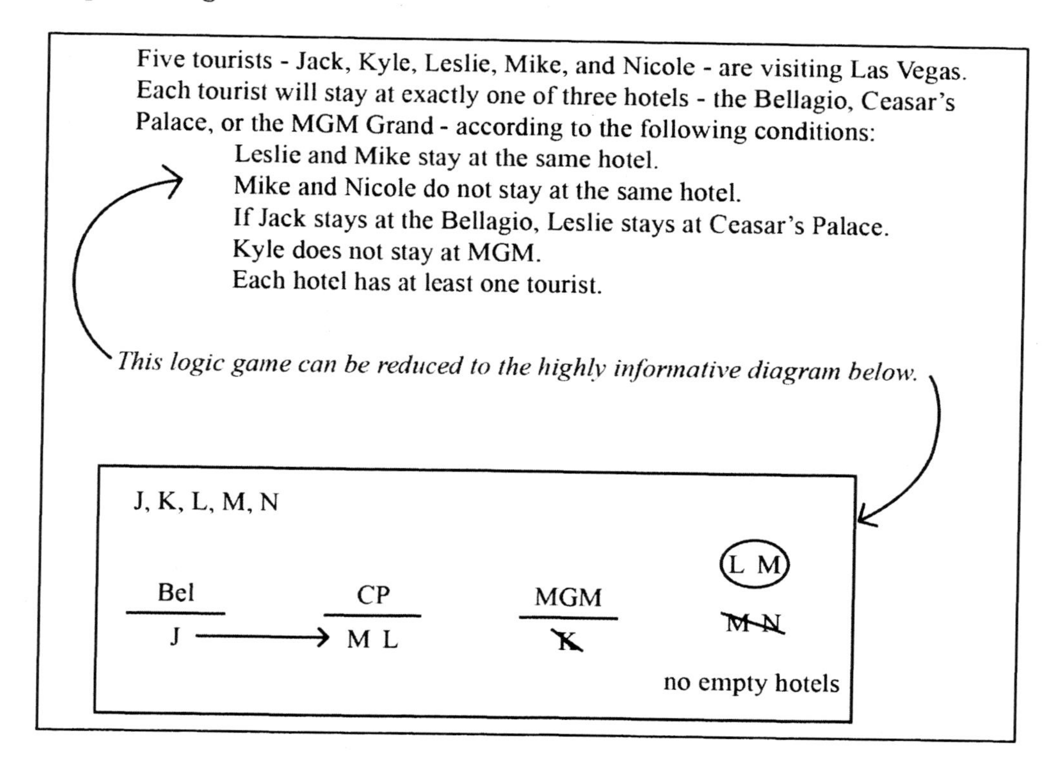

Although diagramming involves a bit of imagination, the lack of artistic abilities will not inhibit your performance. The ability to draw is not being tested on the LSAT. All that is required in diagramming logic games is the ability to translate logical statements into simple symbols. Translating logical statements will be comprehensively discussed, and by the end of this book you will be skilled at diagramming LSAT logic games.

PARTS OF A DIAGRAM

A proper logic game diagram consists of three main parts: (1) the frame; (2) the sidebar; and (3) the work area.

FRAME

The frame of a logic game is the visual structure in which the entities and scenario are drawn. The directive of the game will dictate what the frames look like, and each type of logic game has a distinct frame. Frames for the most common game types are illustrated below.

Example 5.2 - Frames

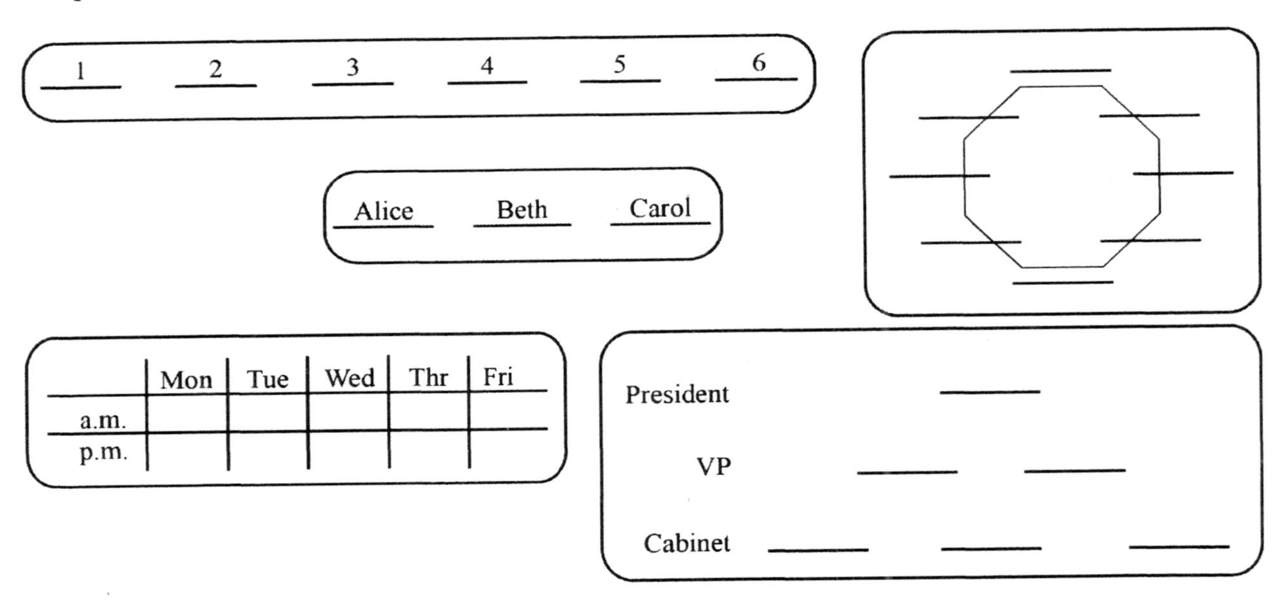

Also see Table 4.2, which provides additional frames associated with common game types.

SIDEBAR

The sidebar is the area of the diagram where relational rules are drawn. Recall that relational rules are those which direct entities relative to other entities. Relational rules do not reference any spots on the frame, so these particular rules cannot be drawn directly into the frame. Accordingly, relational rules are drawn immediately next to the frame in an organized fashion as demonstrated by the following illustration:

Example 5.3 - Sidebar

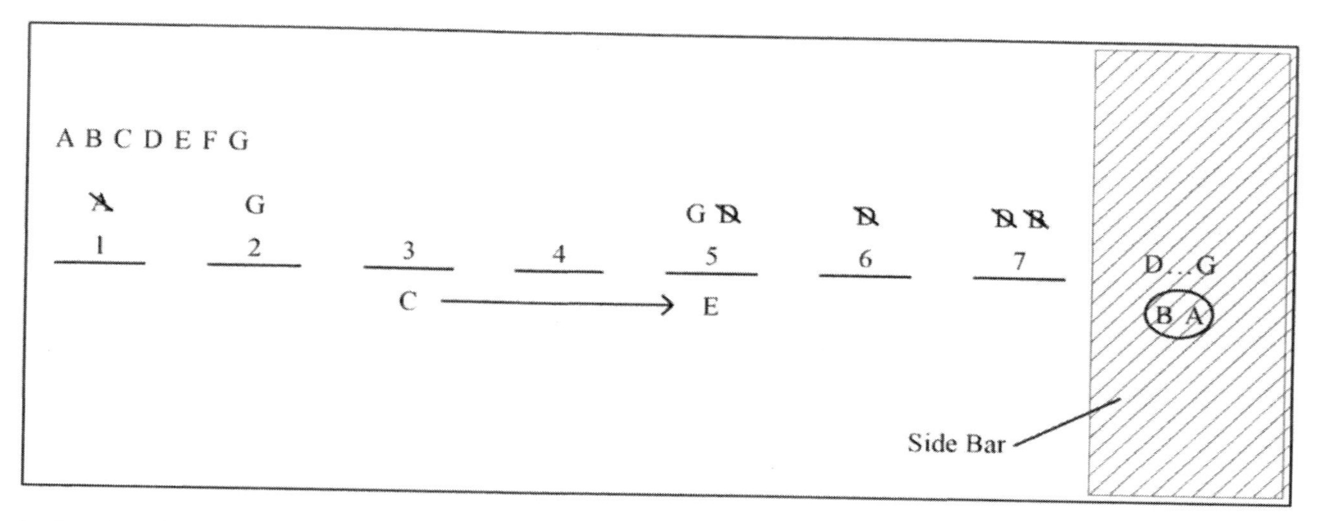

In the above illustration, based on the game in Chapter 3, the rules "Greg shoots sometime after Dennis" and "Adam shoots immediately before Bill," are relational rules and cannot be placed directly into the frame. Unlike the positional rules in this game, these rules are not associated with specific spots. Accordingly, they must be drawn into the diagram in the sidebar.

WORK AREA

The work area is the area on the test booklet that is immediately below the frame. In the work area, you will test new limitations and try different configurations as they are presented in questions.

Example 5.3 – Work Area

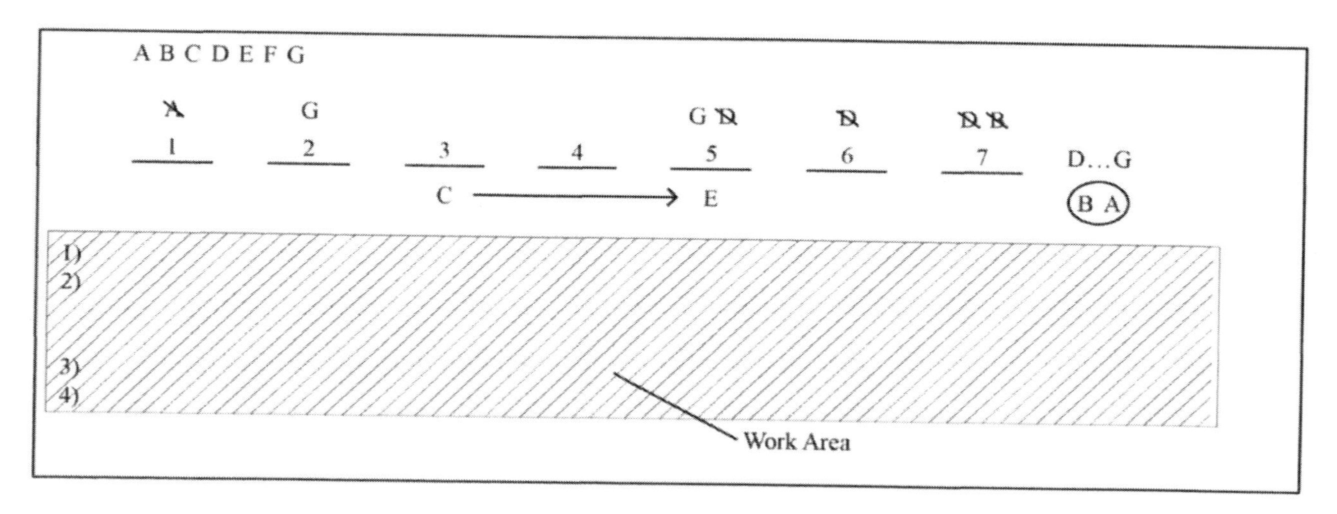

Ideally, you will want to draw the frame in an area of the test booklet that maximizes your work area. Typically, this means drawing the frame in a spot on the page that leaves enough space to work **vertically** down the page, i.e., as high up on the test booklet as possible.

Example 5.4 – Optimal Page Setup

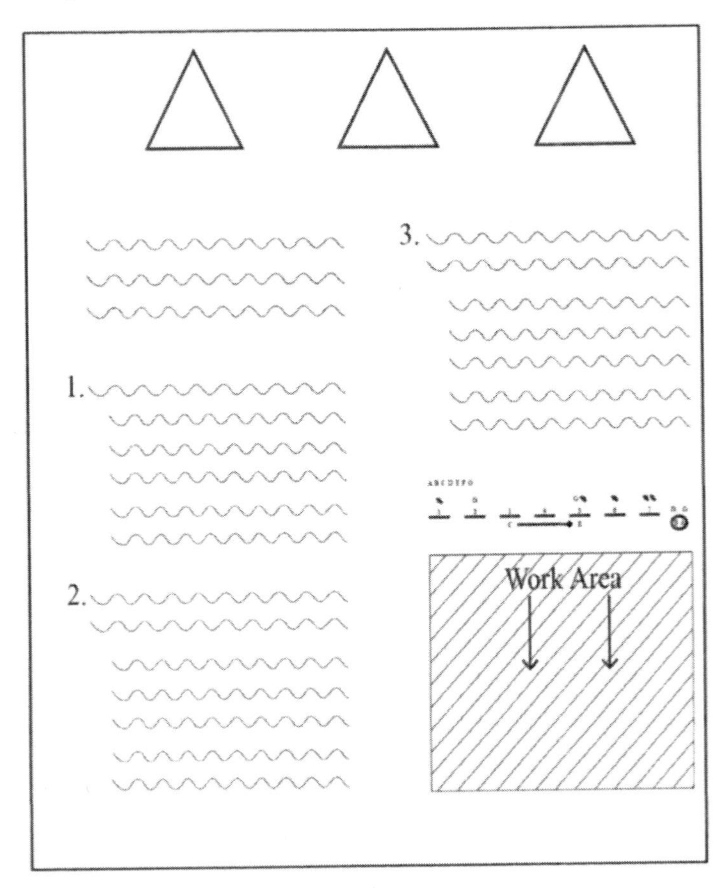

Working vertically when testing answers has several key benefits. Most significantly, it allows you to avoid erasing and re-trying work. Erasing is a no-no when working out logic games. Although you may have an inclination to erase configurations that are not possible (ones that violate rules), do not do this.

Previous work, i.e. work used to solve previous questions, can be beneficial on subsequent questions for several reasons. For one, a configuration that worked without violating any rules may reveal an answer to a subsequent "could be true" question. Further, seeing configurations that do not work may make apparent patterns or deductions that you were not able to catch previously. In order to understand the rules of a logic game, knowing which configurations are not possible can be just as helpful as knowing which configurations are possible. Accordingly, you will want to keep as many tested configurations on your page as possible. Working vertically allows you to do this in an organized manner.

SYMBOLOGY

In order for logic game diagrams to be useful, you must first know how to interpret and implement the associated logical symbols. When you see a symbol in a diagram, it should immediately prompt a logical operation in your mind. The immediate reaction triggered is similar to seeing a stop sign while driving. The red octagon symbol triggers a reaction that tells drivers to brake, even if it cannot be read.

This subsection provides our recommended set of symbols. By using these symbols you can maintain a consistency when drawing diagrams. It is vitally important to be able to quickly draw and interpret your diagrams. Otherwise, there is no point in diagramming. Whether you choose to use our recommended symbols, or your own set, make sure that you stay consistent and develop a routine for using symbols.

HORIZONTAL LINES

Horizontal lines are used exclusively in diagrams to represent *spots*. Each spot is depicted by a single horizontal line, with the name of the spot going ***above*** the line. The spot name goes above the line so that new limitations can be tested underneath. Using horizontal lines facilitates working vertically down a page, which is the preferred way of working on LSAT games. The following example, which is a reproduction of Example 5.2, shows a collection of spots as they would appear in a frame.

Example 5.6 – Spots in a Frame

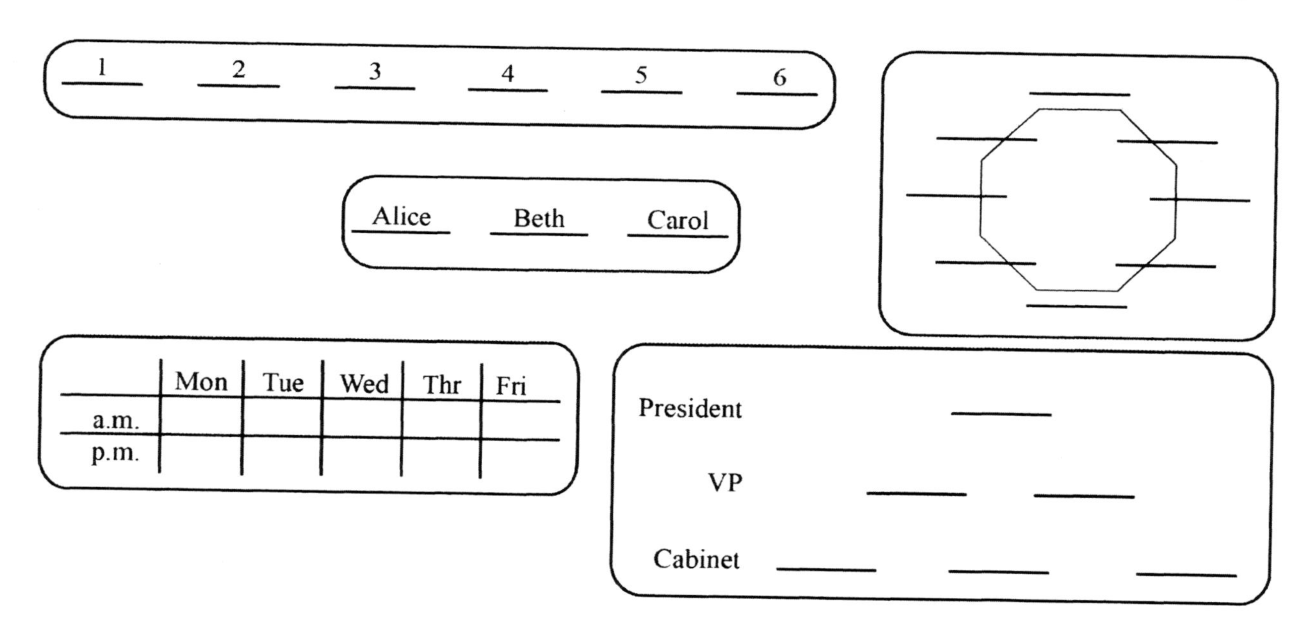

Collectively, the spots are called the *frame;* and each type of game has an associated frame. Recognizing the type of frame to be used in each logic game will save valuable time and earn key points. Table 4.1 is a helpful reference for other common frames. With that being said, we cannot stress enough that it is much more important to understand the Game Directive than to memorize the look of these frames.

CROSSED-OUT ENTITIES

Crossed out entities are used to indicate an impossibility. A crossed out entity in the frame indicates that an entity cannot go in a particular spot.

Example 5.7 – Crossed-Out Entities for Positional Rules

Dennis does not finish third.

1	2	3	4	5	6
		~~D~~			

Dennis finishes in the top four.

1	2	3	4	5	6
				~~D~~	~~D~~

Dennis does not visit Amsterdam.

Paris	Amsterdam	Berlin
	~~D~~	

Two or more entities crossed out in the sidebar indicate that they do not go together in a spot.

Example 5.8– Crossed-Out Entities for Relational Rules

Dennis and Fred do not visit the same city.

~~DF~~

If Dennis is picked to serve on the committee, Fred is not.

~~DF~~ (this is shorthand for D → ~~F~~)

In ordering games, two or more crossed out entities mean that the entities are not in consecutive spots, as pictured.

Example 5.9 – Crossed-Out Entities for Ordering Games

Adam does not sing immediately before Beth.

~~AB~~

The above picture only indicates that Adam cannot be immediately before Beth. However, it *does not indicate* that Adam cannot sing immediately after Beth. Based on this picture, Adam can sing in any spot after Beth, and can sing before Beth as long as it is not immediately before. The examples below show these differences.

Example 5.10 – Clarifying Crossed-Out Entities

Adam does not sing immediately after Beth.

~~BA~~

Adam and Beth do not sing consecutively.

~~AB~~ ~~BA~~

Adam does not sing before Beth.

B...A

Notice in the above example that taking out the modifier "immediately" significantly changes your picture from ~~AB~~ to B...A. Correctly drawing this important difference is absolutely essential.

CIRCLED ENTITIES

Circles are used to indicate definitive placements of entities. Circles in the frame are used to indicate that an entity is permanently associated with a spot. In other words, a circled entity in a spot *always* goes in that spot.

Example 5.11 – Circled Entities for Positional Rules

Zack sees *Gone with the Wind.*

GWW (Z)   Casablanca   Dr. Z

Zack is the last player chosen.

1   2   3   4   5   6 (Z)

Circles in the sidebar are used to indicate that two or more entities are permanently associated with each other, *as drawn*. In games where multiple entities can be placed in a spot, this means that the circled entities in the sidebar will *always* be placed in a spot together.

Example 5.12 – Circled Entities for Relational Rules

Uma and Tom eat at the same restaurant.

Both green cars are in parked in the same lot.

In ordering games, two or more circled entities mean that the entities are in consecutive spots, as pictured.

Example 5.13 – Circled Entities for Ordering Games

Casey sings immediately before Denise.

The above example indicates that Casey sings immediately before Denise only. It *does not indicate* that Casey sings immediately after Denise. Nor does it indicate Casey can sing two, three, four, or more spots before Denise. The examples below show these differences.

Example 5.14 – Clarifying Circled Entities

Casey sings immediately after Denise.

Casey sings before Denise.

C…D

Casey and Denise sing consecutively.

The double arrow above indicates that Casey and Denise are interchangeable. They are consecutive, but it is unknown which one comes first and which comes second.

Also, notice in the above example that taking out the modifier "immediately" significantly changes your picture from CD to C…D. Correctly drawing this important difference is absolutely essential.

UNCIRCLED AND UNCROSSED ENTITIES

The placement of entities into a frame without circles or crosses is used to indicate limited possibilities. An uncircled/uncrossed entity in a frame essentially represents the *inverse* of a crossed out entity in a frame.

Example 5.15 – Positional Rules

Jaime attends history class third period or fifth period.

1	2	3	4	5	6	7
		H		H		

The above rule can also be drawn as follows:

1	2	3	4	5	6	7
~~H~~	~~H~~		~~H~~		~~H~~	~~H~~

However, for purposes of efficiency, the preferred method is to draw the rule using as few symbols as possible. Thus, "Jaime attends history class third period or fifth period" should be drawn with an "H" in spot 3 and 5 rather than the crossed out Hs in the other spots.

ARROWS

Arrows are used in diagrams to illustrate "If-Then" statements. The back end of the arrow is the "If" (trigger). The front end of the arrow is the "Then" (result). Please see Appendix A for more information.

Example 5.16 – Arrows

If Jim presents second, then Beth presents fourth.

1	2	3	4	5	6
	J ⟶	⟶	B		

If Jim visits Amsterdam, then Beth visits Paris.

Paris	Amsterdam	Berlin
B ⟵	⟵ J	

If Jim is on the committee, then Beth is also on the committee.

J ⟶ B

It is important to note that the logic represented by an arrow is unidirectional. J ⟶ B only means "If Jim then Beth" and does not mean "If Beth then Jim." This is why arrows are used, as opposed to straight line connectors.

If Not – Then

To diagram a rule that states "if something happens then something else does not happen" (If X then not Y), the entity on the side of the arrowhead – the result – is crossed out.

Examples 5.17 – If Not – Then

If Jim presents second, then Beth does not present fifth.

1 2 3 4 5 6
J ——————> B̸

If Jim visits Amsterdam, then Beth does not visit Amsterdam.

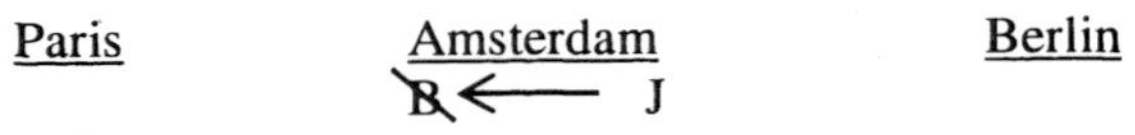

Paris Amsterdam Berlin
B̸ <—— J

If Jim is on the committee, then Beth is not on the committee.

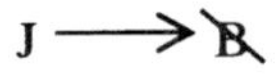

J ——> B̸

However, in matching and selection games J——> B̸ can be simplified to J̸B̸.

Ordering Game Specific Symbols

In addition to the above symbols, ordering games have two additional symbol types: (1) three dots and (2) the blank space. Both of these are used exclusively in the sidebar to indicate relative positions of entities.

Three Dots

Three dots are used to indicate that one entity comes before another. When three dots are used, it is not certain how many spots come between the two entities.

Example 5.18 – Three Dots

Denise sings sometime after Casey.

C…D

It is not certain from the above rule how many spots fall between Casey and Dennis. There may even be no spots in between the two. The only certain information that C…D provides is that there must be a "C" *sometime* before a "D."

Blank Spaces

Blank spaces are used to indicate that a specific number of spots fall between two entities.

Example 5.19 – Blank Spaces

The Ravens finish exactly two places higher than the Pistons.

P __ R

The Pistons finish exactly three spots lower than the Ravens.

P __ __ R

Exactly one team finishes between the Ravens and the Pistons.

R __ P

OBJECTIVES OF DIAGRAMMING

The overarching goal of diagramming on the LSAT is to help you find right answers while saving valuable time. While there are several ways to accomplish this goal, diagramming is the single most effective method. The reason is very simple: in crunch time, illustrated instructions tend to be easier to interpret than written instructions.

There is no denying that pictures aide written instructions and may even replace them. Think about furniture assembly instructions. How much more difficult would assembly be if no illustrations were provided? In GPS systems, why are visual maps provided along with the guiding voice? In patent applications, why do all new inventions require detailed drawings?

Logic game diagrams are no different from these real world examples in that all necessary information is provided through language, but the information can be interpreted and understood more quickly through a visual representation. These visual representations are invaluable tools on the LSAT, and will make your task of solving logic games far less complicated.

The LSAT does not offer much room to work out your thoughts. You will not receive scratch paper or extra materials. Instead, you must learn to use the empty space below the questions and the margins of the paper in your test booklet to work out your answers to the questions. With such a small workspace, it is far more efficient to use symbols and diagrams to solve problems. Specifically, diagrams will guide you to the right answers by helping you to:

1. Organize and visualize thoughts;
2. Recognize and make deductions; and
3. Easily manipulate the game to test answer choices.

ORGANIZE AND VISUALIZE

Diagramming is a useful and convenient way to organize your thoughts before answering questions. When diagramming, you should actively think about the rules and their interplay with the entities. This will allow you to get a feel for the mechanics of the game and absorb the rules.

As you read a logic game in preparation for diagramming, always think *"what is the game telling me with this information,"* as new information is encountered. Look for key words that indicate the type of game and what actions will be performed. For instance, if you see the word "order" you must immediately recognize that the Game Directive is to put things in order, and recall nuances of Ordering Games (i.e., numbered spots, *A...B*, special deductions, etc.). This must be done before drawing any part of the diagram.

When reading the rules, get into the habit of visualizing what they would look like in a diagram. When you see "Adam shoots immediately after Bill," picture this as:

Thus, when diagramming (and also answering questions) you must always see a *BA* somewhere. There will never be an *AB*, *BGA*, *BFA*, *BCGA*, etc. If there is a *B* there will be an *A* right behind it, and you should recognize this immediately.

Example 5.20 – Gathering Information

Carefully read the logic game below and search for the Game Directive: What does the Game want you to do?

> Seven soccer players on a team – Adam, Bill, Carl, Dennis, Ethan, Fred, and Greg – are selected to participate in a shootout. Each player will shoot exactly once and will shoot alone. The order in which the players shoot must be in accordance with the following conditions:
>
> Greg shoots sometime after Dennis.
> Adam shoots immediately after Bill.
> Carl shoots third only if Ethan shoots fifth.
> If Greg does not shoot second, he shoots fifth.

The table below breaks down each piece of information, and what the game is telling you with that information.

Information	Telling you
"Seven soccer players on a team"	7 entities
"Adam, Bill, Carl, Dennis, Ethan, Fred, and Greg"	Entity names: *A, B, C, D, E, F, G*
"Each player will shoot exactly once and will shoot alone"	1 spot for each entity
"The order in which the players must shoot"	**The Game Directive**: Order Entities
"Greg shoots sometime after Dennis."	*D…G*
"Adam shoots immediately after Bill."	(BA)
"Carl shoots third only if Ethan shoots fifth."	If *C* is in spot 3 → *E* is in spot 5
"If Greg does not shoot second, he shoots fifth."	*G* is in either spot 2 or spot 5

Breaking down the information like this will force you to translate the words of the logic game into logical statements, and will allow you to visualize the game. Eventually, you will be able to turn all that information into the following concise and informative diagram:

A B C D E F G

~~A~~	G			G ~~D~~	~~D~~	~~D~~ ~~B~~	
1	2	3	4	5	6	7	D...G
		C	———	→ E			(B A)

MAKE DEDUCTIONS

Deductions are the key to performing well on the LSAT. The ability to make deductions is the primary skill that will separate those who score well on the Games section from those who do not. The ability to deduce information not provided from information already provided is an essential skill that every good lawyer must possess. Accordingly, the LSAT test writers ensure that this ability is thoroughly tested throughout the LSAT, and especially in the Games section.

To this end, diagramming can significantly help with finding deductions. After the initial rules are drawn, a diagram can help combine rules to make deductions. A diagram can also display apparent patterns that lead to deductions. Further, difficult or secondary deductions will become clear when they are drawn and visualized in a diagram.

In the game from Chapter 3, the following deductions are possible.

1. Dennis can never shoot fifth, sixth, or seventh (because Dennis always shoots before Greg, and the latest Greg shoots is fifth).
2. Bill can never shoot seventh (because Adam is always right after Bill).
3. Adam can never shoot first (because Bill is always right before Adam).

The corresponding diagram for these deductions is below. Notice how much easier it is to recall these deductions when they are drawn right into the diagram, as opposed to trying to remember or write them out.

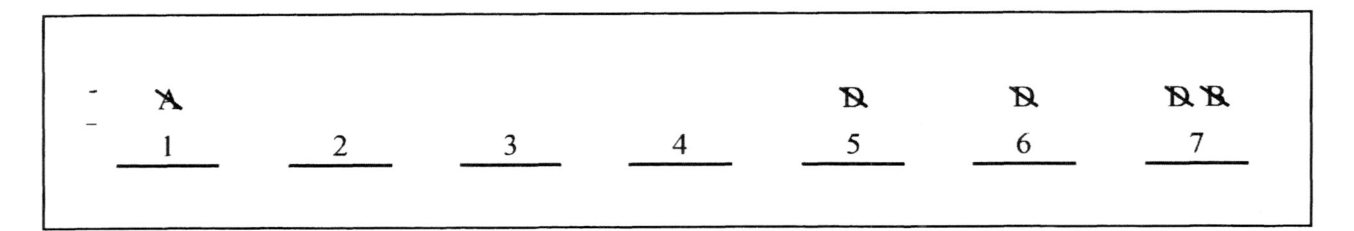

TEST THE ANSWER CHOICES

Unlike in the reading comprehension and logical reasoning sections, some questions in the Games section require a fair amount of work and processing to find the correct answer. Some questions will even require you to "test" the answer choices to see if they are correct.

To test an answer, draw an answer choice into your diagram to determine if any rules are violated. Each answer choice will yield a configuration that may or may not violate any rules. Each answer choice will dictate how the configuration looks, and it is then your job to figure out if any rules are violated.

New limitations are often presented in questions and/or in the answer choices (e.g. If Dennis is fourth, then which of the following could be true?). It is important to note that any limitation presented in a question only applies to that particular question. **You should never carry over a limitation presented in one question over to another question.**

The following example introduces a new limitation through a question and then walks through the method for answering the question.

Example 5.21 – Trying a New Limitation

> Seven soccer players on a team – Adam, Bill, Carl, Dennis, Ethan, Fred, and Greg – are selected to participate in a shootout. Each player will shoot exactly once and will shoot alone. The order in which the players shoot must be in accordance with the following conditions:
>
> Greg shoots sometime after Dennis.
> Adam shoots immediately after Bill.
> Carl shoots third only if Ethan shoots fifth.
> If Greg does not shoot second, he shoots fifth.

New situation that limits this question

4. If Dennis shoots fourth which of the following could be true:
   (A) Greg shoots third
   (B) Carl shoots third
   (C) Bill shoots third
   (D) Bill shoots sixth
   (E) Adam shoots sixth

> *The limitation that Dennis shoots fourth is not a rule.*
> *The limitation applies to this question only.*
> *Dennis may or may not shoot fourth on other questions.*

Question number four presents a new limitation where Dennis shoots fourth. To answer this question, the previously prepared diagram can be used to test the new limitation. The answer and explanation to this question is provided in Chapter 7.

Diagramming is the key to high scores on the Logic Games section. With a neat, small, efficient diagram, you can save time, save space, organize your thoughts, and easily make deductions. We strongly advise that you invest time into learning how to diagram properly.

CHAPTER 6 - DEDUCTIONS

In this Chapter you will learn to:

- Be methodical.
- Scan, combine, and draw.
- Utilize Game Directive information.

THE IMPORTANCE OF DEDUCTIONS

Making deductions may be the single most important task performed when solving logic games. Most test takers will be able understand the given rules to a logic game, and most can draw the rules into a basic barebones diagram. What separates the top scorers from the rest, however, is the ability to manipulate and combine rules to make deductions.

Deductions are made in logic games by combining rules with shared elements to create new rules that further limit the game. In other words, deductions are made by combining two or more rules to make a new rule that encompasses both. You may be thinking, "Why would I want another rule? It's hard enough to keep straight the rules they gave me!" Well, the reason more rules are better, not less, is because *rules limit possibilities*. Possibilities are the enemy in logic games. The fewer possibilities available, the easier it is to place entities and get right answers. The overarching goal in making deductions is to limit possibilities.

STEPS FOR MAKING DEDUCTIONS

Making deductions is a methodical process. If you do not implement a method you can fall prey to second guessing whether key deductions were missed. Further, failing to use a method for making deductions will also slow diagramming, and will waste valuable time. The steps below outline the most effective method to finding key deductions:

1. **Scan** the rules to see if there is an entity or spot common in multiple rules.
2. **Combine** the rules with shared entities or spots, if possible.
3. **Draw** in the new rule.

When following these steps, always keep in mind the ultimate goal is to further limit the game. These steps will be used to make and implement deductions in the following example.

Example 6.1 – Scan, Combine, and Draw

Seven soccer players on a team – Adam, Bill, Carl, Dennis, Ethan, Fred, and Greg – are selected to participate in a shootout. Each player will shoot exactly once and will shoot alone. The order in which the players shoot must be in accordance with the following conditions:

> **Greg** shoots sometime after Dennis.
> Adam shoots immediately after Bill.
> Carl shoots third only if Ethan shoots fifth.
> If **Greg** does not shoot second, he shoots fifth.

SCAN

Notice that **Greg** is common to both the first and last rule:

> "Greg shoots sometime after Dennis."
> "If Greg does not shoot second, he shoots fifth."

This usually means we can combine the two rules to further limit the game.

COMBINE

When combining two rules for a deduction, always begin with the more definite rule. Typically, the most definite rules will be any positional rules that exist. In above example, the rule that has Greg in either spot 2 or spot 5 is positional. ("If Greg does not shoot second, he shoots fifth" is another way of saying he is in 2 or 5):

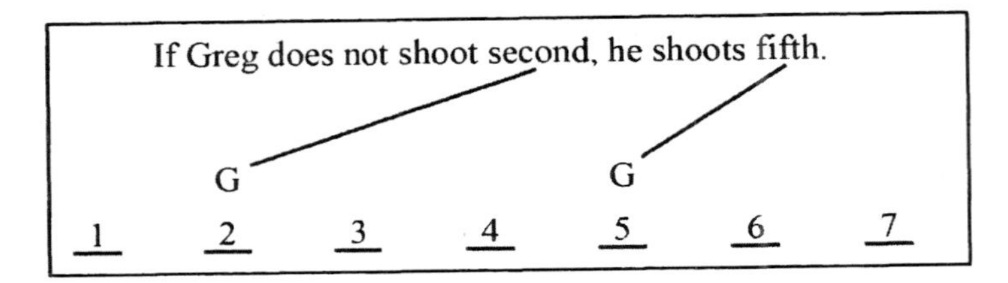

Next, ask: "How does this rule affect the other rule with which it shares an entity?"

The other rule places Greg sometime after Dennis:

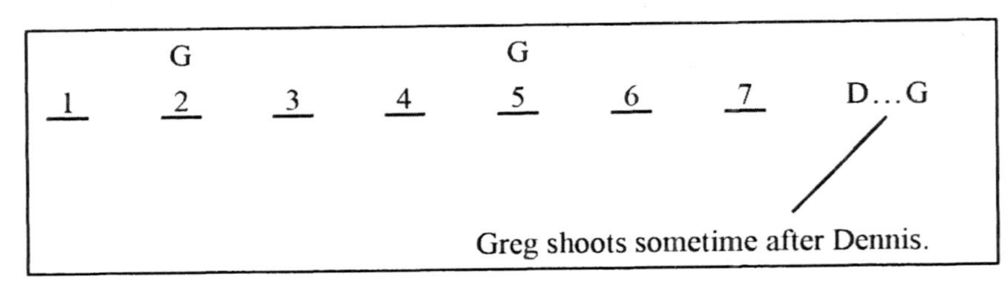

The placement of Greg in spot 2 affects Dennis because Dennis has to shoot before Greg and, therefore, if Greg is in spot 2 Dennis has to be in spot 1:

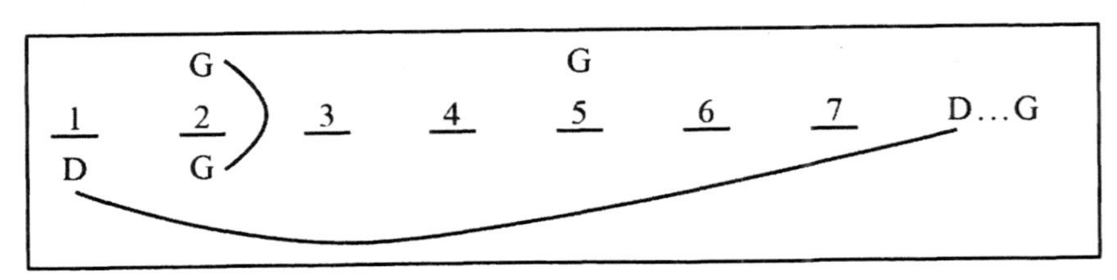

The placement of Greg in spot 5 affects Dennis because Dennis has to shoot before Greg and, therefore, if Greg is in spot 5 Dennis has to be in spots 1, 2, 3, or 4:

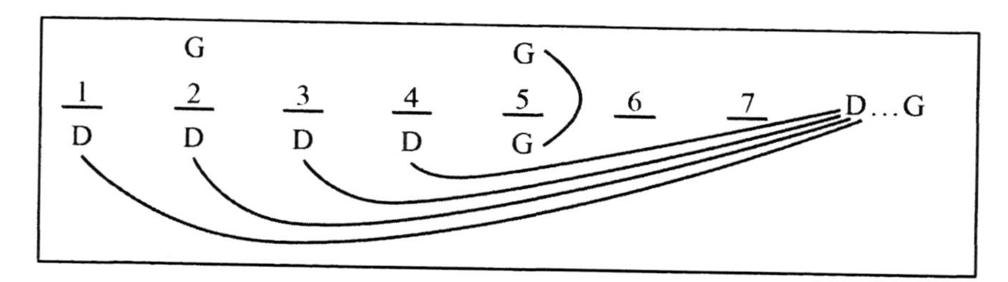

DRAW

What the preceding diagrams show is that if Greg shoots second, Dennis must shoot first (Deduction!), and if Greg shoots fifth, then Dennis must shoot first, second, third, or fourth (Deduction!). When viewed as a whole, you can see that the latest Dennis can ever shoot is fourth. Accordingly, we can make the ultimate deduction that:

DENNIS CAN NEVER BE IN SPOTS 5, 6, OR 7!

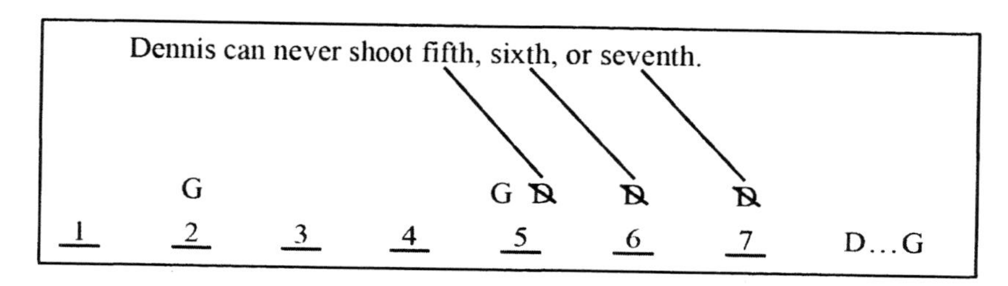

This means any answer choice to a "must be true" or "could be true" question that has Dennis shooting fifth, sixth or seventh is wrong and can be eliminated. Further, if you see Dennis shooting fifth, sixth, or seventh in an answer choice for a "must be false" or "could be true EXCEPT" question, you can be certain that the answer choice is the correct answer. You can be sure that when such a major deduction is available in a logic game it *will* be tested.

GAME TYPE DEDUCTIONS

In addition to the above deductions, a "game type" deduction can be made in this logic game. A game type deduction is a deduction that is specific to the type of game at hand. These deductions can be made when the certain parameters of the game are inherently limiting.

Game type deductions will be discussed in further detail in Chapter 9. For now, just know that, in ordering games, the first and last spots are always special because nothing comes before first and nothing comes after last. If it is known that entity *X* is always before entity *Z* (in your diagram: X...Z), a deduction can be made that *Z* can never be first in the order. Why? Because if *Z* is in spot 1, then *X*, coming before *Z*, would have no spot. The same logic applies to the last spot in an ordering game.

Applying this game type deduction to example 3.1, it can be deduced that Adam can never shoot first and Bill can never shoot last. Why? The second rule states "Adam shoots immediately after Bill." This means if Bill was in spot 7, Adam would have nowhere to go, so this is not possible. Likewise, if Adam was in spot 1, Bill would have nowhere to go, so this is not possible.

SUMMARY OF DEDUCTIONS

To recap, the following deductions were possible in the logic game from example 3.1:

1. Dennis can never shoot fifth, sixth, or seventh (because Dennis always shoots before Greg, and the latest Greg shoots is fifth).
2. Bill can never shoot seventh (because Adam is always right after Bill).
3. Adam can never shoot first (because Bill is always right before Adam).

In a diagram, a summary of the deductions would look like this:

~~A~~				~~D~~	~~D~~	~~D~~ ~~B~~
1	2	3	4	5	6	7

In the next example, an extra rule has been added to the previous logic game. See if you can make any new deductions.

Example 6.2 – New Rule

Seven soccer players on a team – Adam, Bill, Carl, Dennis, Ethan, Fred, and Greg – are selected to participate in a shootout. Each player will shoot exactly once and will shoot alone. The order in which the players shoot must be in accordance with the following conditions:

Greg shoots sometime after Dennis.
Adam shoots immediately after Bill.
Carl shoots third only if Ethan shoots fifth.
If Greg does not shoot second, he shoots fifth.
Adam shoots sixth.

In this example, the added rule requires Adam to shoot sixth. Adam is also involved in the second rule, which requires him to shoot immediately after Bill:

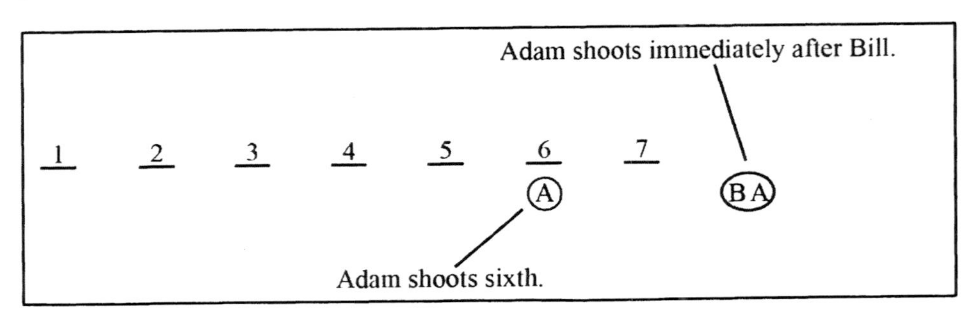

Combining these rules, you should see that Bill must shoot in spot 5 (Deduction!).

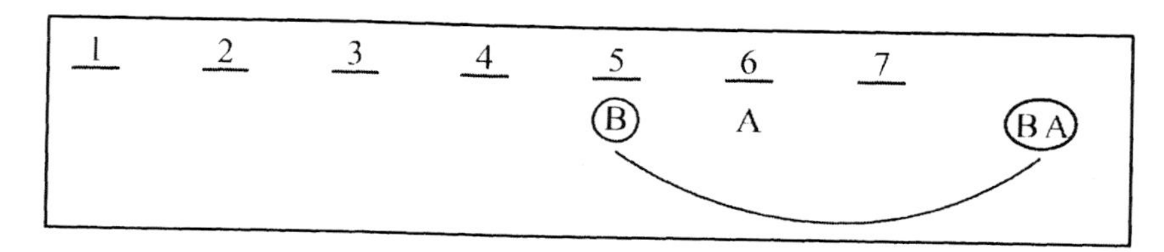

But wait, you are not done yet. So far, only the entities have been checked to see if they appear in multiple rules. The same analysis must also be performed for the spots.

Spot 6, where Adam shoots, is not involved in any other rules, but what about spot 5, where Bill shoots? Spot 5 *does* appear in other rules. In fact, spot 5 is in two other rules:

"Carl shoots third only if Ethan shoots fifth."

"If Greg does not shoot second, he shoots fifth."

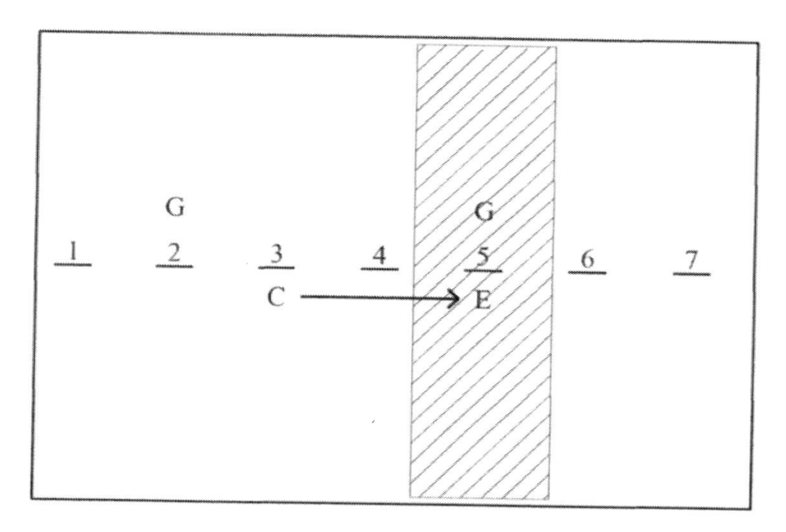

Combining these rules will result in Greg shooting second and Carl blocked off from spot 3:

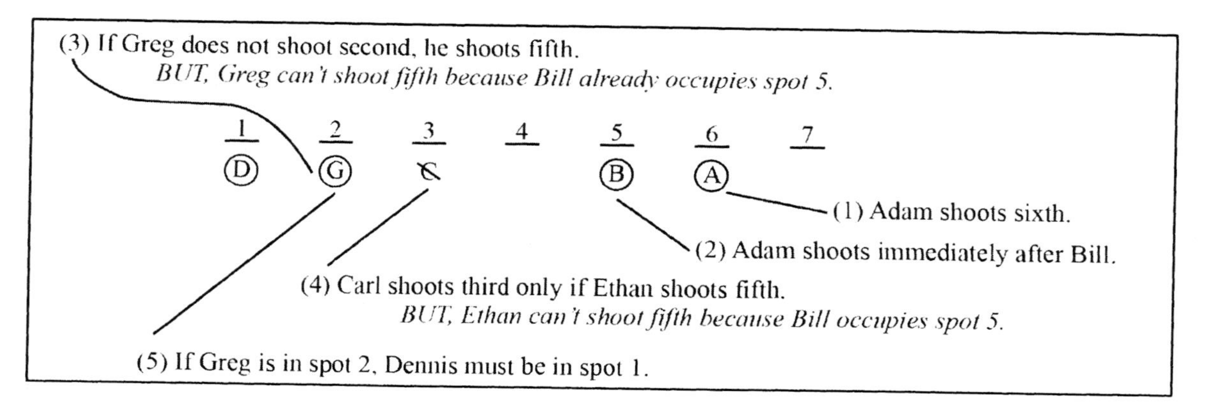

The beauty of the previous diagram is that the rules only locked in one spot – Adam in spot 6. Yet, through deductions and diagramming, it became possible to also lock in Bill and Greg and Dennis and lock out Carl. These deductions significantly reduced the possibilities in the logic game, and gave way to a very informative diagram.

CHAPTER 7 – TESTING ANSWER CHOICES

In this Chapter you will learn how to:

- Test the answer choices to use process of elimination
- Use the answer choices to make further deductions

Once all the necessary deductions have been made in your diagram, it is time to start answering questions and getting points. Each question in the Logic Games section poses new limitations and each answer choice sets forth new configurations to test.

Example 7.1 – Testing the Answer Choices

The following is a typical question on the Logic Games section based on the logic game from Example 3.1:

4. If Dennis shoots fourth, which of the following could be true?
   (A) Greg shoots third.
   (B) Carl shoots third.
   (C) Bill shoots third.
   (D) Bill shoots sixth.
   (E) Adam shoots sixth.

Dennis shooting fourth presents a new limitation that was not part of the background or rules. It only applies to this particular question. Based upon the new limitation, wherein Dennis shoots fourth, five answer choices are presented. Which one of these choices will answer the question correctly?

Often times a deduction would have been made in the diagram that will immediately lead to the correct answer. However, on many occasions you will have to test the possibilities until you find the correct answer. Testing the answer choices will determine which possibilities pass or fail in light of the new limitation and the existing rules. The subsection below demonstrates how answer choices are tested.

START WITH THE NEW LIMITATION

When a new limitation is presented in a question, always begin by drawing into the frame the limitations provided in the question on a new line in the diagram.

If Dennis shoots fourth, begin by placing Dennis into the fourth column:

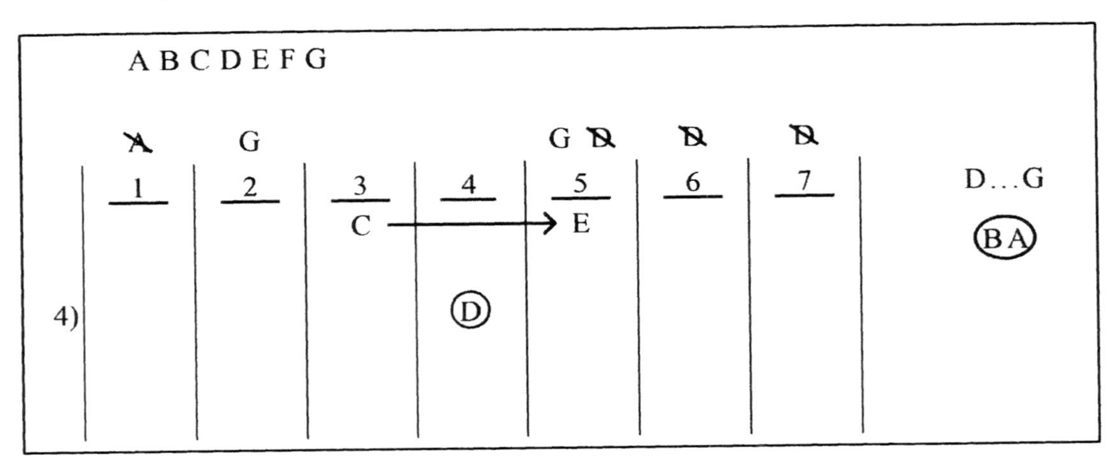

FILL IN DEFINITE ENTITIES

After the limitations from the question are drawn, next draw in definite entities, *i.e.*, entities with known spots. In this particular example, there are no entities with a known spot so you can safely move on.

MAKE FURTHER DEDUCTIONS

Next, check whether the entities or spots in the new limitation are involved in any rules or deductions. If they are, then further deductions can be made. The easiest way to check this is to quickly scan the diagram by looking "up top" to see if the spot in the new limitation has any rules attached. You also need to scan the diagram to see if the entity involved in the new limitation appears anywhere else in the diagram.

Put more simply, after you have drawn the new limitations and the definite entities, ask yourself the following questions:

1. Does the spot involving the new limitation have any other rules attached?
2. Does the entity in the new limitation appear anywhere else in the diagram?

If you answer "yes" to either of these questions, a further deduction can be made. In this example, the answer to the first question is "no." There are no other rules attached to spot 4. When you look to the top of the diagram, spot 4 is clear of any rules. In fact, it is the only spot that does not have a rule attached. Thus, placing an entity in spot 4 does not give rise to a new deduction:

Notice that no rules or deductions were drawn into spot 4.

A B C D E F G

1 2 3 4 5 6 7

C → E

D...G

BA

4)

However, the answer to the second question, "Does the entity in the new limitation appear anywhere else in the diagram," is a resounding yes in this example. For one, the diagram shows Dennis cannot be in spots 5, 6, or 7. This rule has not been violated by the limitation that places Dennis in spot 4, and so no further work is needed.

In addition, Dennis appears in the diagram's sidebar in the rule that states Dennis must shoot sometime before Greg (*D...G*). Thus, *G* must be placed somewhere to the right of *D* on the diagram. Exactly where, however, cannot be determined solely by the rule *D...G*. Thus, you will have to ask whether *G* is involved in any other rules. Greg is involved in one additional rule – he has to shoot second or fifth. Since spot 5 is the only spot to the right of *D* that *G* can be placed in, *G* must go in spot 5:

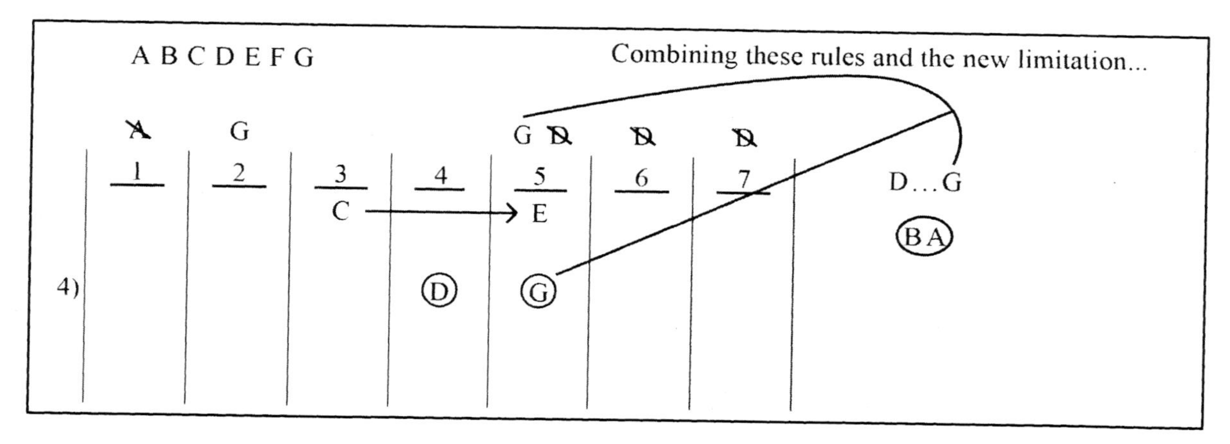

If you look closely at the diagram it should be readily apparent that Greg has to shoot fifth in this situation. This conclusion was far easier to demonstrate with a visual diagram than with words or verbal explanation. This is a quintessential example of why a visual approach to logic games is much more effective than trying to work the problems in your head.

ANSWER THE QUESTION

Now that the new limitation has been diagrammed, answering the question should be fairly straightforward. To refresh, the question and answer choices are:

Example 7.2 – Answering the Question

4. If Dennis shoots fourth, which of the following could be true:

   (A) Greg shoots third
   (B) Carl shoots third
   (C) Bill shoots third
   (D) Bill shoots sixth
   (E) Adam shoots sixth

Right off the bat, you should see that answer choice (A) is incorrect because the diagram clearly shows Greg must shoot fifth:

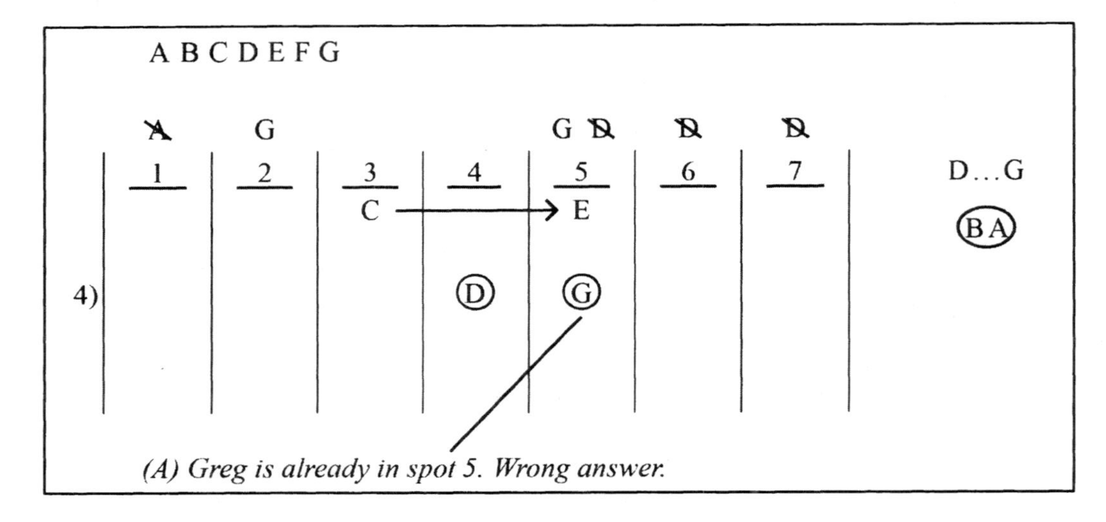

Answer choice (B) has Carl shooting third. Is there anything special about Carl shooting third? Yes, the only rule involving Carl is predicated on him shooting third: If Carl shoots third, then Ethan must shoot fifth. Since Greg is already locked into spot 5, Ethan cannot shoot fifth and Carl cannot shoot third. Therefore, answer choice (B) cannot be true and is wrong:

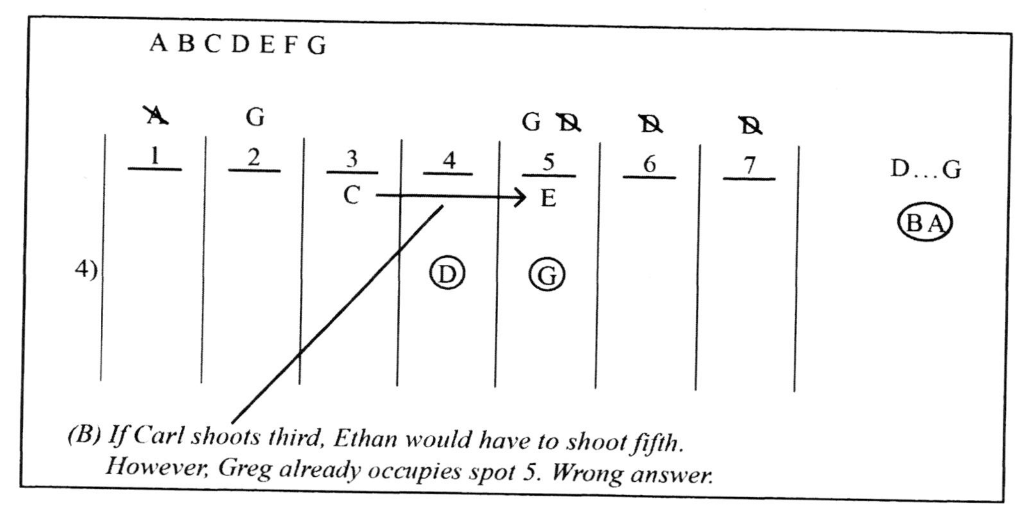

Answer choice (C) has Bill shooting third. Is Bill involved in any rules? Yes, Bill is involved in the rule that requires Adam to follow in the spot immediately after him. Thus, if Bill is in spot 3, Adam would have to shoot in spot 4. However, this is not possible as Dennis is already shooting fourth. Answer choice (C), accordingly, is incorrect:

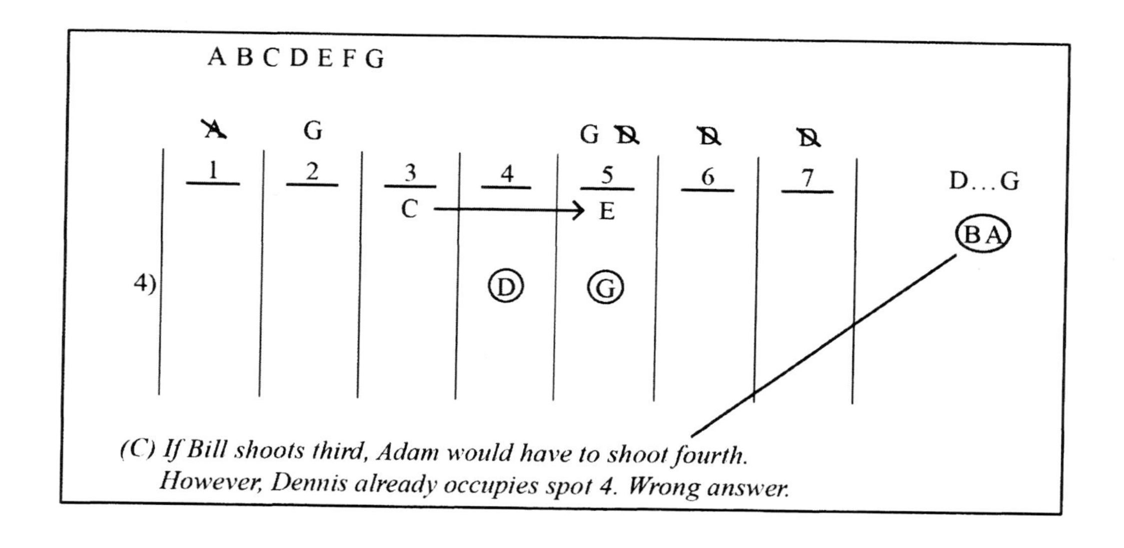

Answer choice (D) is your correct answer. Placing Bill in spot 6 would require Adam to shoot seventh. Spot 7 is available and there are no restrictions for placing Bill in spot 6 and Adam in spot 7:

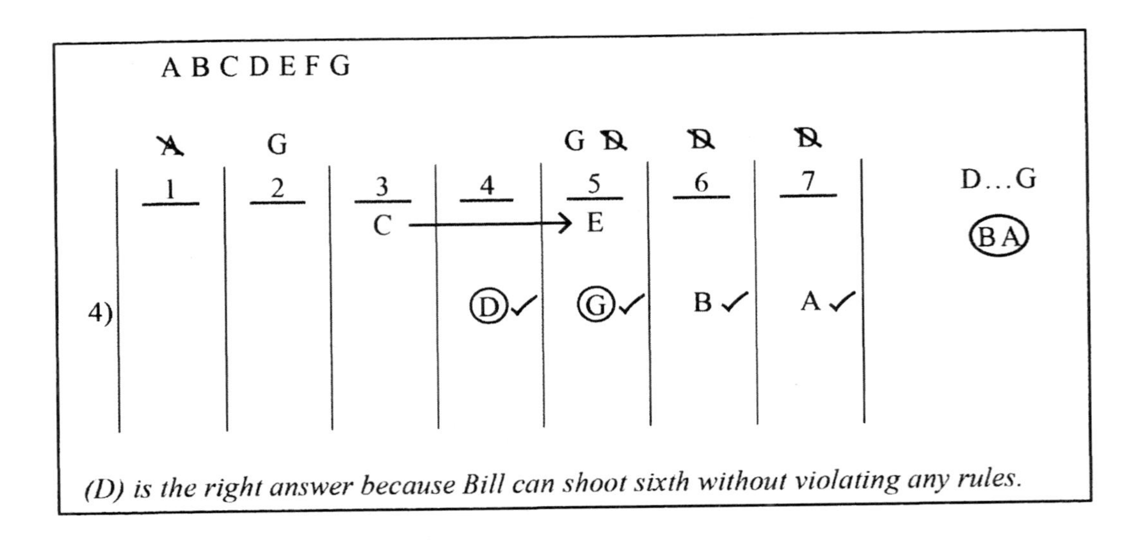

Conversely, answer choice (E) is incorrect, because placing Adam in spot 6 would require Bill to be in spot 5, which is already occupied by Greg:

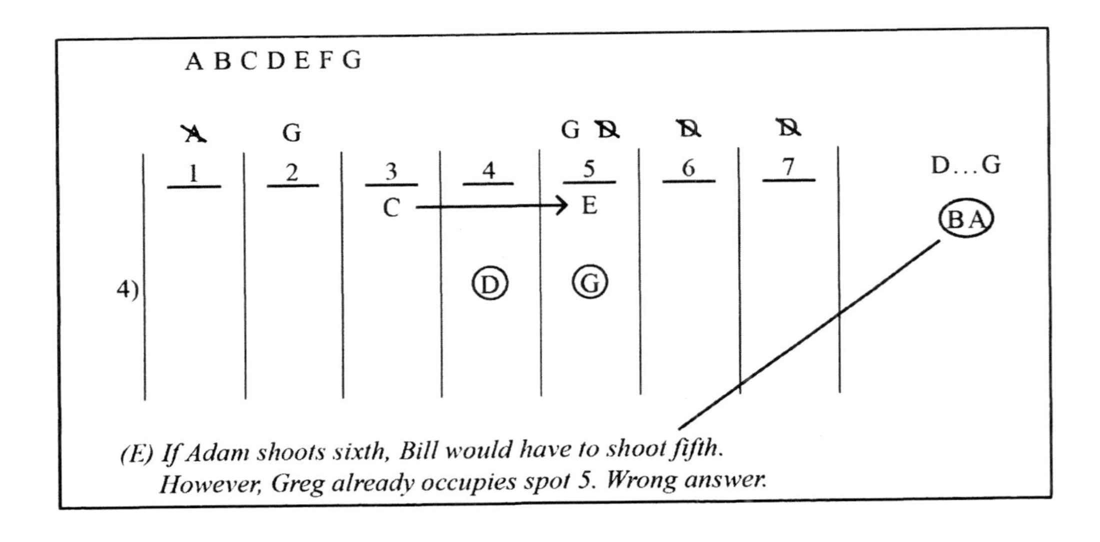

Testing answer choices allows you to eliminate the wrong answers and narrow down the correct answer. Practice will make this process efficient so that you can rapidly narrow your answer choices on test day.

Section III

STRATEGY

CHAPTER 8 – METHOD OF ATTACK

Learn how to read carefully, diagram efficiently, and answer with confidence.

CHAPTER 9 – TIPS FOR EVERY GAME TYPE

Learn tips, shortcuts, and deductions for each common game type.

CHAPTER 10 – PUTTING IT ALL TOGETHER

Learn how to apply your mastery of logic games on test day.

CHAPTER 8 – METHOD OF ATTACK

In this Chapter you will learn to:

- Read carefully.
- Diagram efficiently.
- Answer with confidence.

Now that you familiar with the elements of a logic game and how to diagram, it is time to learn the basic methodology of conquering the vaunted Logic Games section. While other courses will teach five to seven steps to solve a logic game, in truth there are only three fundamental steps required: (1) Read carefully, (2) Diagram efficiently, and (3) Answer with confidence.

While these steps may appear obvious, the *manner* in which you read, diagram, and answer is what makes the difference. The key to this methodology is ensuring that each step is completed with streamlined efficiency, machine-like precision, and keen foresight. Attacking a logic game is both a science and an art, and developing a deep understanding of this method will make working through logic games second nature. Following this method on test day will help you maintain structure, consistency, and confidence throughout the entire Logic Games section.

STEP 1: READ CAREFULLY

The importance of *careful* reading in the Logic Games section cannot be understated. The Logic Games section, unlike the Reading Comprehension and Logical Reasoning sections, has very little "fluff." **Every sentence serves a purpose and is critical to the understanding of a game.**

In the Reading Comprehension section, it is entirely possible to glance over complete sentences in a passage and score well on that passage. As long as you have an understanding of the gist of the passage and know where to find details, you can still perform adequately in the section. In the Logical Reasoning section, if you glance over a sentence or key words in a question, you are not likely to get the question correct; but this error will not carry over to other questions and you could still score well in the section.

Such is not true with the Logic Games section. Simple mistakes such as skipping a rule or glancing over key words will affect your ability to answer *all* questions associated with that game. In other words, not reading carefully in the Logic Games section could easily *cost you an entire game* and approximately six important LSAT points. Recall that six points could mean the difference between your dream school and your safety school! In the Games section, details matter.

WHAT TO LOOK FOR

When reading, begin with the background and *thoroughly* read it to understand the scenario at hand. In reading the background, make sure to identify entities, spots, and the Game Directive. Also identify the game type whenever possible (but do not worry if you are unable to determine the type). After identifying the key components in the background, do the same for rules of the game. Be sure to identify the entities and spots involved in each rule and what is being done to them. Also be sure to recognize any formal logic dispersed in the rules.

STEP 2: DIAGRAM EFFICIENTLY

In order to maximize the value of a diagram, it must be drawn efficiently. Efficient diagramming means capturing as much information as possible, into the most compact and concise diagram that can be drawn. The less you write down on the page, the better; as long as you capture all necessary information.

Efficient diagramming is important for two reasons:

(1) Save time; and

(2) Avoid confusion.

If you efficiently diagram, by only putting in the diagram what is helpful, you can cut valuable seconds off each setup and every question. These seconds will eventually add up and can give you extra time to allocate to harder questions.

Here are some Do's and Don'ts to follow for efficient diagramming:

Do:

- Leave enough room under your diagram to try various situations and possibilities.
- Work vertically.
- Somewhere above the frame, write the entities using abbreviations.
- Use horizontal lines to mark off spots.
- When able, draw rules into the frame of the diagram:
    - Put entities with definite spots into your diagram and circle them.
    - Put in entities that can fall into a limited number of spots directly into the diagram (do not circle).
    - If entity cannot go in a spot, draw it into that spot and *cross it out.*
    - Draw If → Then statements directly into the diagram, when possible.
- If a rule cannot be drawn directly into the frame of the diagram, draw it neatly in the sidebar.

Don't:

- Don't write large. Recall that you have limited space on your test booklet.
- Don't rewrite your diagram or basic frame to try each new situation or possibility (you should be able to use the same master diagram and work vertically off it for each new configuration).
- Don't be redundant. For instance, don't put into a diagram where an entity definitely goes *and* where it cannot go—this would be duplicative.
- Don't write rules in the side bar when they can be placed directly into the frame of the diagram.
- Don't refer back to the written background and written rules once you have drawn the scenario and rules into your diagram. This defeats the purpose of diagramming and wastes precious time.

STEP 3: ANSWER WITH CONFIDENCE

The final step to conquering a logic game is to answer…with confidence. If the methods and tips in this book are judiciously followed, finding correct answers should be straightforward. Once a game is properly diagrammed, and all deductions are made, all of the answers will be right in front of you and you should never have to guess.

Unlike in the Logical Reasoning and Reading Comprehension sections, where you choose the *best* answer, in the Logic Games section you choose the *only* answer. There is no room for interpretation or subjective analysis when answering logic games. Whatever your correctly drawn diagram shows is the right answer will be the right answer. End of story. Thus, as you have read carefully and diagrammed efficiently, you will be able to answer all questions with confidence.

STEP-BY-STEP EXAMPLE – ANSWERING A LOGIC GAME QUESTION

Example 8.1 – Step-by-Step

Seven soccer players on a team – Adam, Bill, Carl, Dennis, Ethan, Fred, and Greg – are selected to participate in a shootout. Each player will shoot exactly once and will shoot alone. The order in which the players shoot must be in accordance with the following conditions:

Greg shoots sometime after Dennis.
Adam shoots immediately after Bill.
Carl shoots third only if Ethan shoots fifth.
If Greg does not shoot second, he shoots fifth.

1. If Adam shoots second, which of the following must be true?

(A) Ben shoots sometime after Carl
(B) Ethan shoots sixth
(C) Greg shoots second
(D) Carl shoots sometime after Dennis
(E) Ethan shoots sometime after Fred

1. Read the Question Carefully

When reading questions, be careful not to overlook or confuse any words. The word "except" is especially significant, as it is commonly used in questions and, if missed, will undoubtedly lead to the wrong answer. Further, a question that asks which answer choice "must be" true should not be confused with a question that asks which answer choice "could be" true. Confusing "must be" and "could be" will lead to wrong answers, so always pay attention to the call of the question. In this example, there is no "except" but there is a "must be," and you must be cognizant of this as you proceed.

2. Identify the Question Type

Table 3.1 lists the various question types you will encounter in the Logic Games section. Familiarizing yourself with this table will help save valuable time on the logic games. Certain questions, such as "acceptability" and some "could be true" questions do not require any written work. Recognizing this can give you extra seconds, even minutes.

This example contains a *must be true* question. Per Table 3.1, the wrong answers could be false or must be false.

3. WHAT IS THE FOCUS OF THE QUESTION?

Start by doing what the question stem tells you to do. Logic games often focus on one entity or spot in the question stem. If an entity or spot is provided in the question stem, your analysis should always begin with that entity or spot. If the question states that Adam shoots second, then put "A" in spot "2" before you do anything else.

In this example, the focus of the question is **Adam** and **spot 2**. Thus, you must immediately put an "A" in spot "2":

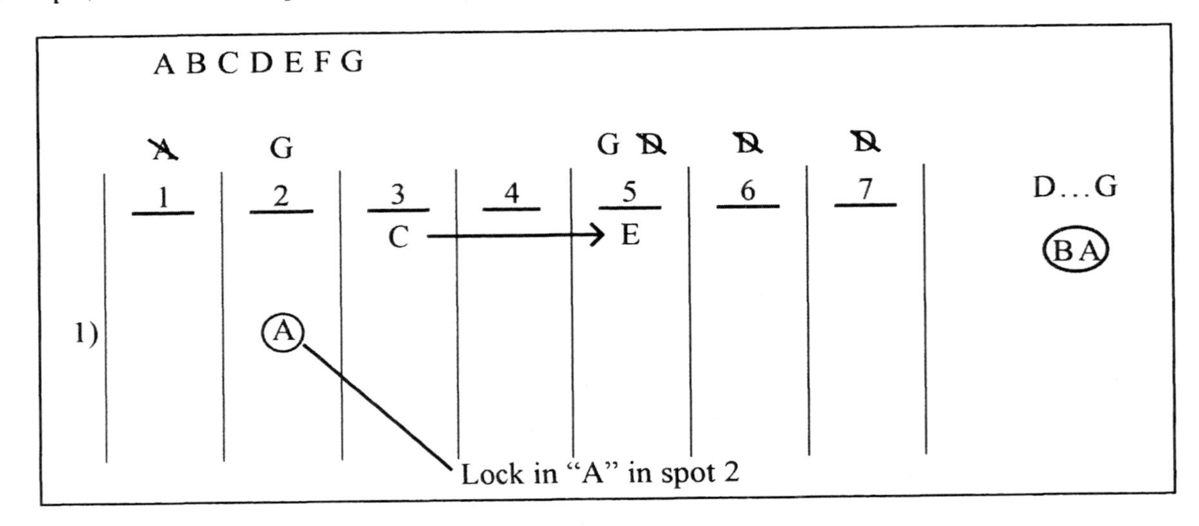

4. LOCK IN WHAT YOU KNOW

If a rule or deduction shows that an entity is definitively in a particular spot (you would have written this down and circled it in your diagram), then for each question that requires work you should immediately put that entity into its assigned spot before proceeding. For instance, if a rule says "Zack always visits Detroit," in every situation attempted you must draw "Z" in spot "D" before proceeding.

In this example, the rules do not indicate that any entities can be locked-in. Thus, step four is satisfied and you can move on to step five.

5. WHICH RULES ARE INVOLVED?

After you have identified the question type, recognized what the question is asking of you, and locked-in the initial entities, the next step is to make deductions, as discussed previously. The question stem in this example asks "If Adam shoots second, which of the following must be true." We know the second rule, which states "Adam shoots immediately after Bill," is involved. Why? Because Adam is the focus of this question and Adam is involved in the second rule. In other words, *you see Adam in both the question and rule two.* Once we identify a rule is involved in a question, we then follow the rule and lock entities into our diagram.

In this example, it is known that “A” goes in spot “2.” So, which rules involve “A” and spot “2?” Rules two and four.

Rule two says: “*Adam* shoots immediately after Bill.” Rule four says: “If Greg does not shoot *second*, he shoots fifth.” Thus, these two rules are implicated by the question stem and are going to play a key role in answering the question.

Rule two requires “B” to always come immediately before “A.” Accordingly, knowing “A” has to be in spot 2 (per the question stem), “B” can definitively be locked in spot “1”:

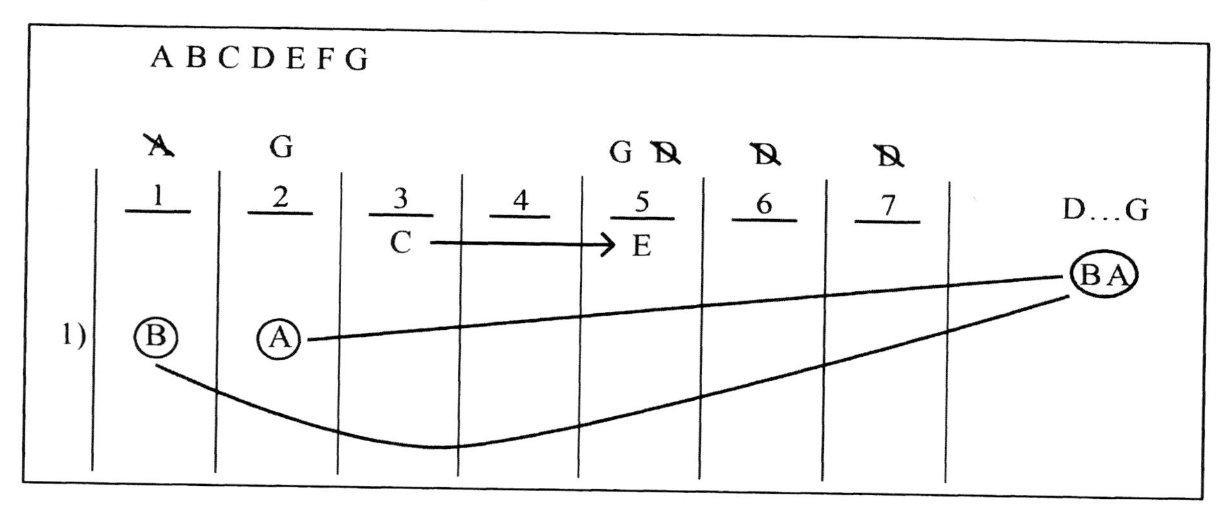

Rule four requires “G” to be in spot “2” or spot “5.” Since spot “2” is already taken by “A” (per the question stem), then “G” has to go in spot “5”:

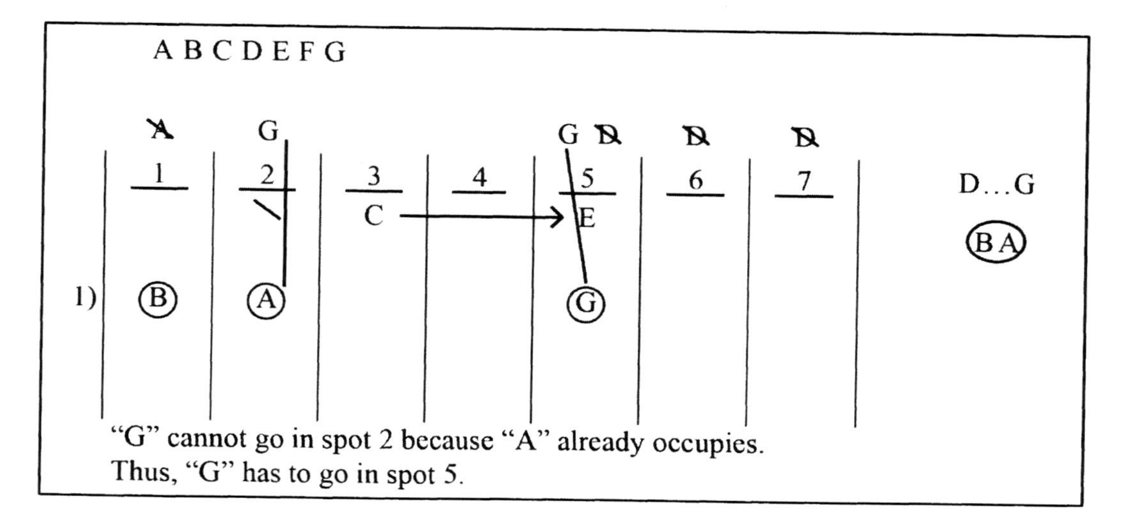

Thus, there are now three entities locked-in, which significantly narrows possibilities.

We are now at the critical horizon. Those who stop at or before this point will score in the 25th percentile on the LSAT, and those who continue on can reach the 75th percentile (or higher). Every time you lock in an entity/spot into your diagram you must go back to the rules and see which rule *that last* entity/spot implicates. Once you find another rule that is implicated you follow that rule and lock in more entities, and repeat this until there are no more rules implicated.

In this example, you just put “B” in spot “1” and “G” in spot “5”; therefore you have to see if “B”, “G”, spot “1”, or spot “5” are involved in any rules. Per the rules, neither “B” nor spot “1” are involved in any rules, and therefore you are done with those entities. Conversely, both “G” and “spot “5” are involved in other rules, and therefore, will need more attention.

“G” is involved in rule one: “Greg shoots sometime after Dennis.” This means, per the diagram, that “D” has to be in spots “3” or “4” because it cannot be after “G”, and spots “1” and “2” are already taken by “B” and “A”:

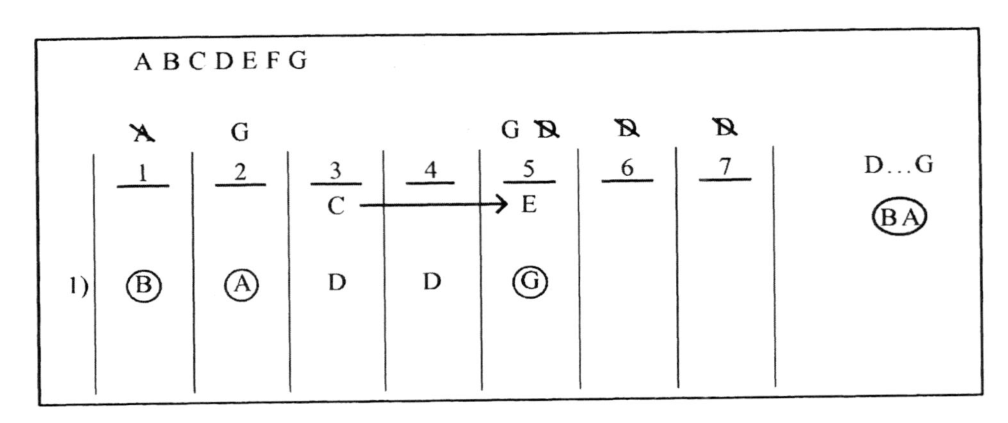

Spot “5” is involved in rule three, as the spot on the backside of the arrow of the if-then statement: “Carl shoots third only if Ethan shoots fifth.” As previously determined, the rule is merely stating if “C” is in spot “3” then “E” would have to be in spot “5.” But, since we know that “G” is already in spot “5” something is amiss.

Can this if-then statement provided work under the current conditions? In this example, the if-then statement cannot work because “E” cannot go in spot “5” since “G” is already in spot “5.” Thus, we know that “C” cannot go in spot “3” because, from rule three, it would force “E” into the already occupied spot “5”:

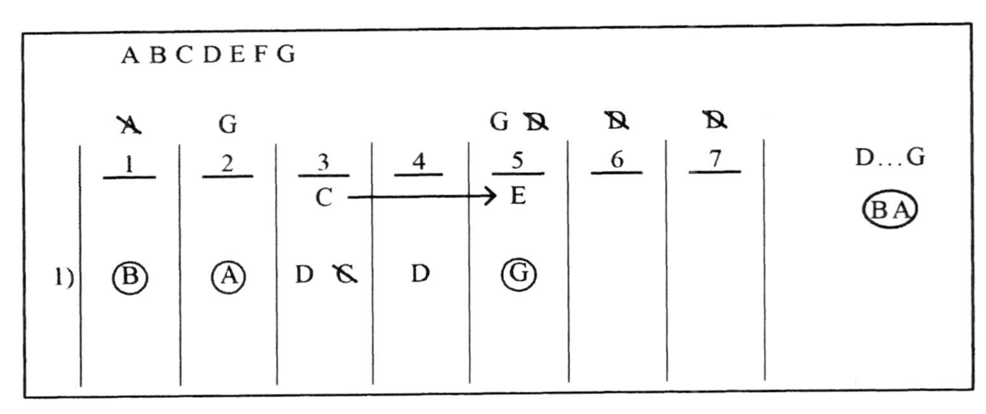

Thus we can conclude from our diagram all of the following:

1. Adam shoots second.
2. Bill shoots first.
3. Dennis shoots either third or fourth.
4. Carl cannot shoot third.
5. Greg shoots fifth.

6. Cross off incorrect answers

Any answer choice that conflicts with the diagram or conflicts with your work is wrong and can be crossed out. Let's go through our example and cross off answers that conflict with our diagram:

1. If Adam shoots second, which of the following must be true?

(A) Ben shoots sometime after Carl – We can cross this answer off because it conflicts with our diagram. Ben is in spot one and does not shoot after anyone, including Carl.

(B) Ethan shoots sixth – This answer does not conflict with our diagram so we cannot immediately cross it off. However, from Table 3.1, we know that answers that "could be false" are wrong in "must be true" questions. Ethan *could* shoot sixth, but he can also shoot third, fourth, or seventh. So while it could be true that Ethan shoots sixth, this also could be false. Thus, this is a wrong answer and we can eliminate it as well.

(C) Greg shoots second – We can cross this answer off because it conflicts with our diagram. Greg has been placed in spot 5 and circled. Accordingly, we know Greg can only shoot fifth and, therefore, cannot shoot second.

(D) Carl shoots sometime after Dennis – This answer does not conflict with our diagram, so we cannot immediately cross it off. What do we know about Carl and Dennis? We know Carl has to be in spots *4, 6,* or *7.* We know Dennis must go in either spots *3* or *4*. Therefore, we know Carol has to be sometime after Dennis. **This is your correct answer!**

To be sure, on "must be true" questions you can ask yourself "Is there any way that the answer choice can be false?" If there is no way it can be false, then you know you have the correct answer.

In this example, there is nowhere we can place Carl so that he is before Dennis. The earliest we can place Carl is in spot *4*. But if we place Carl in *4* then Dennis would have to go in spot *3*, since Dennis can only go in *3* or *4*.

(E) Ethan shoots sometime after Fred – This answer does not conflict with our diagram, so we cannot immediately cross it off. What do we know about Ethan and Fred? Not much really. We know they cannot go in spots *1, 2,* or *5* since those spots are taken by Bill, Adam, and Greg respectively. Other than that, Ethan and Fred can pretty much go anywhere. Thus, we cannot state for certain whether Ethan shoots sometime after Fred. This could be true or could be false, and based on Table 3.1, we know that this answer, therefore, is wrong.

7. Circle the correct answer

Your very last step should be to circle the correct answer, and then bubble it in your answer sheet…with confidence.

USING OLD WORK

On occasion, you will be able to use your prior work to answer a question. This occurs most often with "could be true" questions. When you encounter a "could be true" question that has an answer you have tried before and did not violate any rules, you know this is the correct answer. No work needs to be done.

Example 8.2 – Using Old Work:

Question 1 is the same as from our previous example and we will use our previous work to answer question 2.

1. If Adam shoots second, which of the following must be true?
   (A) Ben shoots sometime after Carl
   (B) Ethan shoots sixth
   (C) Greg shoots second
   (D) Carl shoots sometime after Dennis
   (E) Ethan shoots sometime after Fred

2. If Greg shoots fifth, which of the following could be true?
   (A) Adam, Ethan, and Fred shoot consecutively
   (B) Dennis shoots sixth
   (C) Ben shoots first
   (D) Ethan, Carl, and Ben shoot consecutively
   (E) Carl shoots third

Right off the bat, this seems like it is going to be a time consuming question. Answer choices (A) and (D) involve three entities each and do not tell you where those entities go. Typically you would have to use trial and error to test each answer choice. This could consume a ton of time and brain power.

However, there is a much faster way to get the correct answer. Use your old work. Question #1 had Ben in spot 1, Adam in spot 2, and Greg in spot 5. The other entities could go in various other spots in order to make the configuration work. Thus, we knew that a scenario with B in spot 1, A in spot 2 and G in spot 5 is possible.

How does that help us here? Well, if we put Greg in spot 5 like question two tells us, it looks somewhat like the situation from Question #1, which was a possibility that passed. Scanning the answers, we see that one of our fixed entities from Question #1 is present. Answer choice (C) states "Ben shoots first." We know this can be true because we saw it was true in the previous question when Greg shot fifth. Thus, without doing any work for Question #2, we can confidently choose (C) as our answer!

CHAPTER 9 – TIPS FOR EVERY GAME TYPE

In this Chapter, you will learn:

- Tips and shortcuts for each common game type.
- Possible deductions based on common game type.

The information, steps, and tips learned in the previous Sections are applicable to all logic games, in general. These methods should be broadly applied to each and every logic game encountered. If you have a good handle on the basic methodology expressed and understand the Game Directive, you should be well on your way to achieving the games score you desire.

With that being said, there are certain shortcuts and tricks available for each type of game that can help you shave extra time and score more points. This section details the shortcuts and tricks for some of the more common game types.

GAME DIRECTIVE: ORDER ENTITIES

ASSIGNED ORDERING

Assigned Ordering games require you to place entities in a set order based on the limitations provided in the rules. These games are extremely common on the LSAT. In fact, you should expect to see *at least* one ordering game on the exam (either Assigned or Relative), and more often than not there will be one Assigned Ordering game.

EXAMPLE 9.1

Seven basketball teams – Jaguars, Kings, Lions, Mavericks, Nighthawks, Owls, and Pistons – will be seeded 1 through 7 for a tournament. The seeding must follow the following conditions:

The Lions are seeded third
The Owls are seeded exactly two spots lower than the Kings
The Mavericks are not seeded sixth or seventh
If the Jaguars are seeded first the Owls are seeded fourth

FRAME

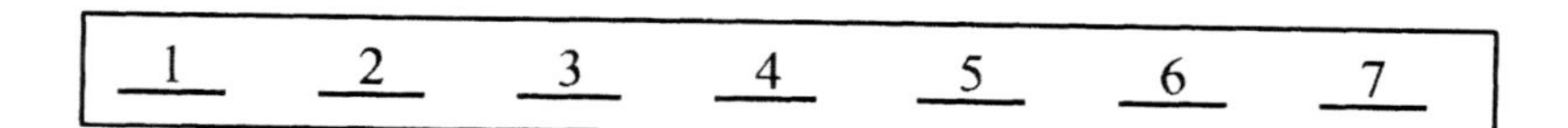

GAME TYPE DEDUCTIONS AND TIPS

In Assigned Ordering games, the spots on both ends of the frame, the first and last spots, tend to be especially significant. When rules dictate that one entity is always placed sometime before another entity, end spots will be blocked off for one of those two entities.

In the preceding example, the Owls are seeded lower than the Kings. This means that the Owls can never be in the very last spot and that the Kings can never be in the very first spot. In this specific example, the Owls are exactly two seeds lower than the Kings, so the Owls are actually blocked off from spots 6 *and* 7. Likewise, the Kings are blocked off from spots 1 *and* 2:

1	2	3	4	5	6	7
~~K~~	~~K~~				~~O~~	~~O~~

This game type deduction is further illustrated below:

Rule: X comes sometime before Y. **Symbol:** X...Y

1	2	3	4	5	6	7
~~Y~~						~~X~~

Rule: A is exactly three spots before B. **Symbol:** A __ __ B

1	2	3	4	5	6	7
~~B~~	~~B~~	~~B~~		~~A~~	~~A~~	~~A~~

RELATIVE ORDERING

In Relative Ordering games, entities are placed in an order, relative to each other. Relative Ordering games are different from assigned ordering because there are no set spots as in most other games.

EXAMPLE 9.2

Six college football teams – Kansas, LSU, Michigan, Nebraska, Oregon, and Pittsburgh – are being ranked in a national poll of top teams from one to six, with one being the "highest" rank and six being the "lowest" rank. The teams are ranked according to the following rules:

Michigan is ranked higher than Oregon
Pittsburgh is ranked higher than Kansas but lower than Michigan
Nebraska is ranked lower than Kansas and LSU

FRAME

M…O

M…P…K

K…N

L…N

GAME TYPE DEDUCTIONS AND TIPS

In Relative Ordering games, rules will specify how entities fit into games relative to other entities only (using Relational Rules). Accordingly, there is no real frame, but rather, the diagram is comprised of just entities separated by dots to show relative positions. Special deductions can almost always be made in relative ordering games by connecting two rules if they share common entities.

In this example, M…O and M…P…K both have *M* in common. Accordingly, you can combine these two chains to create one chain.

We do not know whether O goes before or after P and K. We only know that M is ranked higher than O, P, and K. Accordingly, the chain looks like this:

```
    ... O
M
    ... P... K
```

Likewise, K…N and M…P…K both have *K* in common. Accordingly, you can add K…N to the chain as follows:

```
    ... O
M
    ... P... K... N
```

Lastly, L…N and K…N can also be connected:

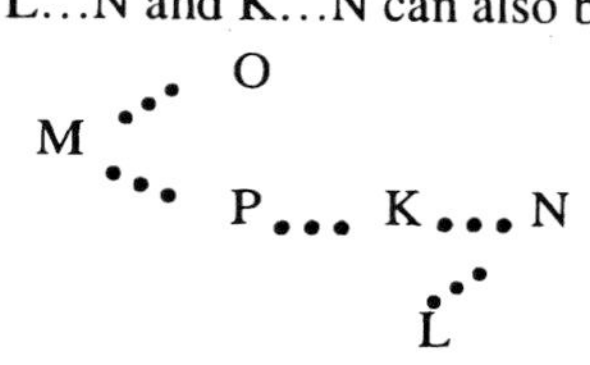

Thus, by combining the rules, we have one big chain that contains a great deal of information that is easy to see and utilize. **From this chain, we can importantly deduce the only entities that can come first and the only ones that can go last.** Only entities that do not have a branch to the left can come first. In the above example, only Michigan and LSU can be ranked number 1. Conversely, only entities that do not have a branch to the right can go last. In the preceding example, only Nebraska and Oregon can be last. These deductions will greatly aide you in answering questions.

GAME DIRECTIVE: MATCH ENTITIES

MATCHING

Matching games require entities to be matched with spots based on the limitations provided in the rules. Typically, each spot can hold multiple entities. Matching games are very common on the LSAT and some form of a matching game should be expected.

EXAMPLE 9.3

Three veteran cops – Alan, Beth, and Carrie – and four rookie cops – Quinn, Rick, Sam, and Tina – are being assigned to traffic duty. The cops will be assigned to one of three posts – Downtown, Uptown, and Midtown, -- according to the following conditions:

Each rookie cop must be assigned to a post with a veteran cop.
Sam is not assigned to the same post as Rick.
If Beth is assigned to Uptown then Tina is assigned to Uptown also.
Carrie is assigned to either Downtown or Midtown.
Each post has at least one, but no more than three, cops.

FRAME

Downtown	Uptown	Midtown

GAME TYPE DEDUCTIONS AND TIPS

In matching games, spots that have identical limitations and entities that have identical limitations are special and will provide hidden deductions. These deductions will be immensely helpful when answering questions, especially in must be true questions.

Downtown	Uptown	Midtown	
	B → T ~~C~~		~~SR~~

In the above example, the Downtown and Midtown spots have identical limitations attached (no limitations to be exact). We can use this fact to easily eliminate answers in the following question.

If Quinn and Rick work Uptown, which of the following must be true:

(A) Beth works Downtown.
(B) Beth works Midtown.
(C) Alan works Uptown.
(D) Beth works Uptown.
(E) Carrie works Midtown.

In this question the answer choices that involve Downtown and Midtown can easily be eliminated, as those two spots are equivalent and there is no way of knowing exactly which entities go there based on the information provided. Answer choice (A), (B), and (E) can be eliminated immediately, leaving you with a 50/50 chance of getting this question correct without having to do any work. The correct answer is (C) because Alan is the only veteran can work Uptown, since Carrie can never work Uptown, and Beth cannot work Uptown if there are already two cops Uptown (because if Beth is Uptown, Tina must also be Uptown).

SELECTION

Selection games require you to select entities from a larger group of entities. There is no frame for Selection games.

EXAMPLE 9.4

Venezuela Records, a music record label, has given instruction to a talent scout to sign several new artists to diversify the label's collection of artists. The talent scout scoped out two country artists - Amerigo and Billy Bob Cyrus – two rappers – Crunk Master and DJ Logic – three rock bands – the Eaglers, Foghorn, and the Great to Be Alive, -- and three pop singers – Hart, Iglesias Brothers, and Jets. The talent scout is required to sign no less than three and no more than five artists based on the following restrictions:

Only one rapper can be signed
If Hart is signed the Eaglers cannot be signed
If Foghorn is signed the Great to Be Alive must be signed
If a country artist is signed at least one rapper and one pop singer must be signed
If no country artists are signed the Eaglers must be signed
If Crunk Master is signed then Amerigo must be signed along with either Jets or Iglesias Brothers

1. Which is a complete and accurate list of artists that can be signed together

(A) DJ Logic, Foghorn, Great to Be Alive, and Amerigo
(B) Crunk Master, Amerigo, Jets, and Billy Bob Cyrus
(C) Amerigo, DJ Logic, Eaglers, and Iglesias Brothers
(D) Crunk Master, Billy Bob Cyrus, Amerigo, DJ Logic, Hart
(E) Amerigo, Crunk Master, Foghorn, Great to Be Alive

FRAME

Selection games, generally, have no frames and will only have relational rules (like relative ordering games).

In this game, the diagram would look as follows:

H → ~~E~~
F → G
coun → rap
coun → E
C → AJ or AI

GAME TYPE DEDUCTIONS AND TIPS

Selection games are comprised of several if-then statements. Thus, deductions can always be made by combining the if-then statements. To do this, see if any two of the if-then statements have any common entities – with one being on the back side of the arrow and one being on the front side of the arrow.

Based on the preceding diagram, it is not readily apparent that two if-then statements can be combined. However, recall that the first rule $H \not\rightarrow E$ is a "not" rule (if H then not E), for which a contrapositive can be drawn. See Appendix A for a detailed discussion on contrapositives. The contrapositive of $H \rightarrow \not{E}$ is $E \rightarrow \not{H}$. With $E \rightarrow \not{H}$ we can now combine this rule with coun → E as both rules have an E on opposite sides of the arrow. Making this combination we now have the following deduction:

$$\text{coun} \rightarrow E \rightarrow \not{H}$$

Amerigo and Billy Bob Cyrus are the only two country artists, so the following ultimate deductions can be made:

$$A \rightarrow E \rightarrow \not{H}$$
$$B \rightarrow E \rightarrow \not{H}$$

OR

$$A \text{ or } B \rightarrow E \rightarrow \not{H}$$

In other words, if a country artist is signed then Hart cannot be signed. *Specifically, if Amerigo or Billy Bob Cyrus are signed then Hart cannot be signed.* These deductions will be immensely helpful when answering questions.

GAME DIRECTIVE: POSITION ENTITIES

POSITIONING

Positioning games require you to place entities around a particular shape pursuant to the given rules. They are somewhat like ordering games, except nonlinear.

EXAMPLE 9.5

A square-shaped mall food court has exactly eight restaurants: two taco joints – Taco Bell and Del Taco – three burger joints – McDonalds, Burger King, and Wendy's – and three ethnic cuisines – Panda Express, La Shish, and Sbarro's. The restaurants are arranged two to a side according to the following constraints:

The north side of the food court has at least one burger joint.
Panda Express and La Shish are not on the same side of the food court.
The two taco joints are on opposite sides of the food court.
Wendy's is on west side of the food court.
Taco Bell is on a side that is adjacent to the side on which Sbarro's is located.

FRAME

In positioning games, it is absolutely imperative that you draw the correct shape as your frame. Depending on the scenario, the frames can be just about any shape; although shapes with an even number of sides (octagon, hexagon, and square) are the most common.

In above game, the shape of the mall food court is provided directly in the first sentence. However, often times the shape of your frame will have to be inferred from the background or rules. For instance, if the background states that six people are seated evenly spaced around a table, you know that your frame is a hexagon.

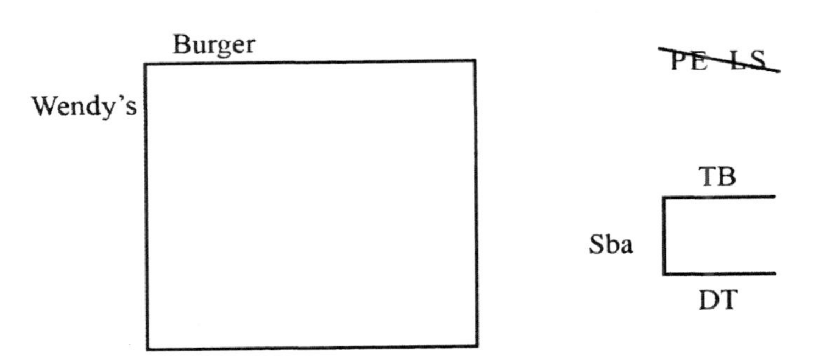

GAME TYPE DEDUCTIONS AND TIPS

They keys to positioning games are twofold. First, draw the correct shape. Second, determine as many adjacent and opposite relationships between entities as possible. The preceding diagram is illustrative of three key points:

1. In positioning games there is no "working down" on the page. You will have to draw your diagrams to be as small as possible. If you run out of space, you may have to erase and redraw your work.

2. Notice the deduction that can be made wherein Del Taco has to be adjacent to Sbarro's. This will help to significantly limit possibilities of where entities can be placed, especially if Sbarro's is on the north or south side (as this would leave the entire west side blocked off by Wendy's and a taco joint).

3. The three-sided figure with TB, Sba, and DT *does not* state that Taco Bell is always on the north side, Sbarro's is always on the west side, and Del Taco is always on the south side. To indicate that, you would have drawn the same figure, but have it circled. When it is uncircled, the figure simply signifies the relative positions of Taco Bell, Sbarro's, and Del Taco to each other.

MAPPING

Mapping games require you place entities on a map based on a given set of rules.

EXAMPLE 9.6

The state of Petoria has five counties – North Washington, South Washington, Jackson, East Jackson, and Lincoln. North Washington is entirely north of South Washington. Jackson is entirely east of North Washington and South Washington. East Jackson is entirely east of Jackson. Lincoln is entirely north of Jackson, entirely west of North Washington, and entirely west of East Jackson. Eight bus stops are to be placed in each of these counties according to the following rules:

Lincoln has at least two bus stops.
East Jackson has no more than one bus stop.
If North Washington has more than one bus stop then South Washington does not have any bus stops.
If Jackson has three or more bus stops then Lincoln has exactly two bus stops.

FRAME

The below frame is basically what your frame should look like. There are no constraints on the dimensions of the counties so the frame can look quite different and still contain the correct information.

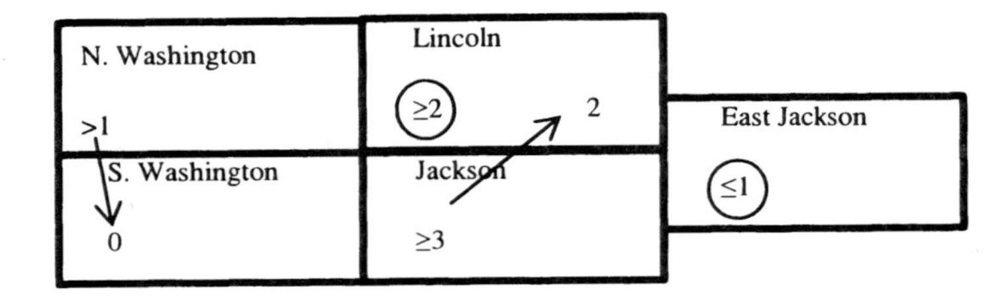

GAME TYPE DEDUCTIONS AND TIPS

Mapping games are fairly straightforward once the correct maps are drawn. Further, these games are not common on the LSAT. If encountered on the exam, take time to precisely draw the map. If you have the map drawn exactly at is described, the questions will not be very difficult.

CHAPTER 10 – PUTTING IT ALL TOGETHER

In this Chapter, you will learn:

- Micro Strategy: Method for attacking the Games.
- Macro Strategy: Long term plan for LSAT preparation.

MICRO STRATEGY

While we do not advocate relying upon a rigid step-by-step method, it is helpful to develop time-saving habits described below.

1. Quickly flip through your Games section to see the lineup

Games Sections can vary in difficulty, but the level of difficulty will usually average out among the games. If you encounter one extremely difficult game, you will probably see a very easy game in the same section. You also may encounter four moderately difficult games. You may have two easier games, and two harder games. Scanning the section for easier games can benefit you by allowing you to attack easy games first. In turn, this will allow you to get easy points and allot extra time for the harder games.

2. Attack the logic games in a deliberate order

Ideally, you will want to start with the easiest logic game and progress to the hardest one. Since each question is worth the same amount of points, you want to try to get as many points with the easy games first. Factors to consider include the length of the game, the type of the game, the number of rules, and the number of questions.

Shorter games are often easier to interpret. Games types that are easy to recognize, such as Assigned Ordering games, tend to be more straightforward. Games with more rules will limit possibilities, possibly lending to an easier game. Finally, you will want to contemplate the number of questions each game has when determining your order of attack. A game with more questions will yield more points, and so this should also be considered.

Unfortunately, it is not always apparent which game in the section is the easiest, and which is the hardest, until you actually complete the games. Some logic games are verbose and appear difficult, but may be simple once you make deductions. Predicting the difficulty level of games under time constraints is not an exact science. The more you practice, the better your instincts will be.

Thus, it is recommended that you attack the *first* manageable game you encounter. If you always do well on matching games, try starting with the first matching game you encounter. If you are strong at all game types, perhaps find the logic game with the most questions and start there. However you decide to attack the section, ensure that you maximize your strengths.

For more information, see Chapters 4 and 9.

3. Quickly Scan the Entire Game

a. Background: Identify Entities and Scenario

b. Rules: Positional and Relational

Before you start diagramming and answering questions, you must have an accurate grasp on the game as a whole. Scanning the background will help you identify the entities and understand the scenario, including the Game Directive and game type. Scanning the rules will help you determine the nuances and difficulty of the game. The quick scan will put you in prime position to understand the big picture of the game so that you can accurately diagram and eventually answer questions.

For more information, see Chapter 3.

4. Isolate the Game Directive

Identify the Game Directive: What is the game asking you to do? Does the game want you to order entities, match entities, or position entities? Or does the game want you to do a combination of those things?

For more information, see Chapter 4.

5. Identify Game Type *if possible*

After you identify the Game Directive, identify the game type *if possible*. If the game is a common game type, you will know how to proceed. Move forward with your knowledge of the correct frame and game type deductions.

However, if you cannot immediately identify the game type or you are not sure, do not fret. Sometimes under the pressure and time constraints of test day, you may find yourself unable to determine an exact game type. This is a common occurrence, so do not panic. Focus on your method of attack and the Game Directive to break down the logic in the game and create useful diagram. Once diagrammed, you can answer the questions in a systematic way.

For more information, see Chapters 4 and 9.

6. Carefully Read the Background and Rules

After you understand the Game Directive and the big picture, you must read the background and rules carefully so you do not miss or misread any words. Recall that every word in a logic game is important.

For more information, see Chapter 8.

7. Diagram

a. Draw your frame and list your entities.

b. Draw positional rules within the frame.

c. Place relational rules in your sidebar.

d. Make deductions and draw them into your frame and sidebar.

e. Ensure sure you have adequate workspace to test your answer choices

For more information, see Chapters 5 and 6.

8. Answer the Questions with Confidence

a. Acceptability questions first

Always start with acceptability questions, if present. Acceptability questions are not only easy, but they also allow you to check whether you have accounted for all the rules introduced in the game. Acceptability questions such as "Which of the following can be an accurate schedule of classes from Monday through Friday?" will require you to go through each rule introduced and see if any of the answer choices violate that rule. In doing so, you will find four configurations that do not work and one configuration that does. This information can then be used to answer subsequent questions.

b. Easy before difficult

Every question is worth one point on the LSAT, so get easy points before you attempt difficult questions that consume more time. However, remember that you are the one taking the test, and your subjective sense of what is easy or difficult matters. Do the questions you find to be easiest first. Table 3.1 provides a list of questions in their general order of difficulty.

c. Specific before general

Specific questions – ones that introduce a new limitation in the question – should be tried before more general questions for two reasons. For one, they tend to be easier. Second, specific questions can provide answers to subsequent general questions. An example of a specific question is: "If X happens, which of the following must be true?" This type of question can typically be answered with less work, and can provide a working configuration to help answer a subsequent question.

Conversely, general questions tend to be more difficult and require more work to answer outright. An example of a general question is: "Which of the following could be true?" Without using prior work or tested configurations, general questions can be extremely time consuming.

For more information, see Chapters 3 and 8.

9. Not in Master Diagram? Then Test the Answer Choices

Some questions can be answered solely from the master diagram. However, others will require you to test each answer choice. In doing so, work efficiently in your workspace and eliminate as many wrong answers as possible. If you have diagrammed properly and understood the question, you will find the only correct answer.

See Chapter 7 for more information.

MACRO STRATEGY: BEFORE TEST DAY

UNDERSTAND YOUR PREFERENCES, STRENGTHS, AND WEAKNESSES

In order to use your time efficiently, you must understand where you stand. Be realistic and identify your weaknesses and strengths early in your study preparation. As famed football coach Knute Rockne once stated, to be successful you must "Build up your weaknesses until they become your strong points." You must increase accuracy on your weak game types and increase speed on your strong games types. Do not fall into the trap of gaining false confidence from practicing only games with which you are comfortable. Push yourself until you are able to master the very hard games in fair time.

LEARN TO MANAGE YOUR TIME

On the Logic Games section, you have 35 minutes to solve 4 logic games and answer 22 to 24 questions. That averages to about 8.75 minutes a game and approximately 1.5 minutes per question. Considering that you must carefully read the background and rules, and create an accurate diagram, you actually have much less than 1.5 minutes per question. In light of this daunting time crunch, it is essential that you learn to manage your time well before test day.

To do this, you must practice time-saving methods and put yourself on stringent time limitations when practicing. For one, you must be able attack and solve easy logic games within five to six minutes. Even if you are comfortable with easy games, do not neglect them when studying. Focus not only on gaining easy points with the simple games, but also focus on gaining time with these games. The faster you can solve an easy game, the more time you will have to allocate to the harder games.

Additionally, it is imperative to instill a timing regiment when practicing a section. After your initial few weeks of practice, you must never allow yourself to exceed 35 minutes on a section. The ultimate goal is to be able to finish the Logic Games section in 32 minutes, so that you can leave yourself a three minute cushion. The cushion can allow you to:

1. Check answers on which you were unsure;
2. Answer questions you skipped due to their difficulty;
3. Take your time on a logic game that is particularly difficult; and
4. Compose yourself if you begin to panic.

In order to increase your speed, we suggest that you start by allowing yourself 35 minutes on the section and then slowly reduce the amount of time you allow yourself, until you are comfortably finishing your practice tests within 32 minutes. Your brain will begin to adjust, and become accustomed to the shorter times as you train yourself to practice with less and less time.

SLOW DOWN TO SPEED UP

If you, like most students, are caught in a bind between trying to get more questions correct on your section and trying to go faster, try to first slow down and focus on accuracy. Start practicing Games Sections and allow yourself 50 minutes instead of 35. Work methodically and consistently and focus on getting all the answers correct. Do several games sections with 50 minutes in this manner.

The following week, cut down to 45 minutes for a Games Section. The week after, cut to 40 minutes. Eventually, get to 35 minutes. Finally, in the weeks before test day shave down to 32 minutes. The idea is to give your brain a little space to perform more accurately and comfortably. Once you have increased your accuracy, speed up. You shave down your times gradually and condition yourself to perform comfortably in the shorter time period.

The above is just an example, but the principle can be beneficial for many students at different paces and levels of accuracy. If you are a student who struggles to finish a Games section in 35 minutes, give yourself space. Keep track of your time but let yourself work through a few games sections until you finish. Then cut several minutes from that time and work games sections

in your new shorter time. Then shave several minutes off that time, and so forth. Slow down for accuracy; then speed up gradually.

BE METHODICAL, NOT MECHANICAL

The rules, tips, and strategies in this book are meant to help you. Do not simply memorize them and engage in fruitless repetition. Understand them, internalize them, and engage with the material to develop your best method.

BE NEAT

Not everyone is blessed with beautiful penmanship or the ability to draw clean lines. Nevertheless, neat and careful diagramming is a skill that can be learned, and for students taking the LSAT it *must* be learned. If you waste space on your LSAT test booklet, you will run out of room to work out the correct answers to questions. If you have sloppy diagrams, you can confuse your thought process and lose time and precious points.

Practice drawing neat, small diagrams that are space efficient on your practice tests. Your work need not be a masterpiece, but it should be as neat as possible and serve as a tool to keep your thoughts organized. Again, if you must slow down a little to improve your diagramming, do it. Then speed up again gradually.

PRACTICE, PRACTICE, PRACTICE

Practice makes perfect. Practice will help you learn to manage your time, choose correct answers, eliminate wrong answers, learn your strengths and weaknesses, diagram efficiently, and trust your instincts.

Section IV

PRACTICE

GAMES PART 1

Very easy to moderately difficult logic games. Get used to answering the questions quickly and accurately.

GAMES PART 2

Complex logic games. Allow a few extra minutes for these games.

SOLUTIONS

Detailed answers with diagrams and explanations.

Games

PART 1

Biff's Workout

Biff is strength training to get in shape for swimsuit season. Biff will work out six areas of his body – shoulders, chest, back, abs, legs, and glutes. Biff's workout consists of a three day cycle. He will work out two different areas of his body each day. In addition, he will run on a treadmill every other day. The training regimen will be scheduled according to the following rules:

Biff will work out his back on the same day as either his shoulders or his glutes.
The day Biff works out his abs is also a day when he will run on the treadmill.
The day Biff works his chest cannot be a day that he runs.

1. If Biff runs on day 1, which of the following must be false?
   (A) Biff works his glutes on day 1.
   (B) Biff works his glutes on day 2.
   (C) Biff works his glutes on day 3.
   (D) Biff works his back on day 1.
   (E) Biff works his back on day 2.

2. If Biff works his glutes on day 2, then which of the following could be true?
   (A) Biff works his glutes with his back.
   (B) Biff works his glutes with his shoulders.
   (C) Biff works his glutes with his chest.
   (D) Biff works his abs with his shoulders.
   (E) Biff works his chest with his shoulders.

3. If Biff works his chest and his legs on different days, then which of the following must be true?
   (A) On the day Biff works his legs, he also runs.
   (B) On the day Biff works his back, he also runs.
   (C) On the day Biff works his glutes, he also runs.
   (D) On the day Biff works his glutes, he does not run.
   (E) On the day Biff works his shoulders, he does not run.

4. Which of the following must be false?
   (A) Biff works his chest on day 1 and his abs on day 2.
   (B) Biff works his abs on day 1 and his chest on day 2.
   (C) Biff works his chest on day 1 and his abs on day 3.
   (D) Biff works his chest on day 2 and his abs on day 3.
   (E) Biff works his abs on day 2 and his chest on day 3.

5. If Biff works his back on day 1 and runs on day 2, then which of the following must be false?
   (A) Biff works abs and legs on the same day.
   (B) Biff works abs and shoulders on the same day.
   (C) Biff works glutes and abs on the same day.
   (D) Biff works glutes and back on the same day.
   (E) Biff works glutes and legs on the same day.

6. Which of the following is an acceptable schedule?
   (A) day 1: shoulders, glutes
   day 2: abs, legs, run
   day 3: chest, back
   (B) day 1: abs, legs, run
   day 2: chest, shoulders
   day 3: back, glutes
   (C) day 1: abs, back, run
   day 2: shoulders, chest
   day 3: glutes, legs, run
   (D) day 1: back, shoulders
   day 2: abs, glutes, run
   day 3: chest, legs
   (E) day 1: chest, glutes, run
   day 2: back, shoulders
   day 3: abs, legs, run

Scheduling Patients

A psychologist is scheduling appointments for seven patients – Sam, Tom, Ulf, Victor, Walt, Xavi, and Yasser. The psychologist plans to work from 9am to 4pm, with a one hour lunch break from noon to 1pm. The patients will be scheduled for 1 hour appointments according to the following restrictions:

Sam does not have the 9am or 10am time slots.
Ulf and Xavi are seen consecutively.
Walt has an appointment at some time before Victor.
Yasser has an appointment at either 4pm or 11am.

1. Which is an acceptable order for the appointments from 9am to 4pm?
   (A) Walt, Sam, Victor, Ulf, Xavi, Tom, Yasser
   (B) Walt, Victor, Ulf, Xavi, Yasser, Sam, Tom
   (C) Tom, Walt, Yasser, Xavi, Sam, Ulf, Victor
   (D) Tom, Victor, Sam, Ulf, Xavi, Walt, Yasser
   (E) Walt, Xavi, Ulf, Victor, Tom, Sam, Yasser

2. If Tom has the 11am appointment, which of the following could be true?
   (A) Ulf's appointment is at 1pm and Sam's appointment is at 2pm.
   (B) Walt's appointment is at 1pm and Victor's appointment is at 2pm.
   (C) Ulf's appointment is at 10am and Walt's appointment is at 3pm.
   (D) Sam's appointment is at 3pm and Yasser's appointment is at 1pm.
   (E) Walt's appointment is at 1pm and Victor's is at 4pm.

3. Each of the following could be true EXCEPT:
   (A) Walt's appointment is at 11am and Ulf's appointment is at 1pm.
   (B) Walt's appointment is at 9am and Ulf's appointment is at 10am.
   (C) Walt's appointment is at 9am and Ulf's appointment is at 11am.
   (D) Walt's appointment is at 11am and Ulf's appointment is at 10am.
   (E) Walt's appointment is at 10am and Ulf's appointment is at 11am.

4. If Ulf has an appointment exactly one hour before Victor, which must be true:
   (A) Walt has an appointment sometime before Tom.
   (B) Walt has an appointment sometime before Sam.
   (C) Tom has an appointment sometime before Walt.
   (D) Tom has an appointment sometime before Sam.
   (E) Sam has an appointment sometime before Tom.

5. If Walt has the 11am appointment, which is a complete and accurate list of patients that could have the 2pm appointment?
   (A) Ulf, Xavi, Victor, Tom, Sam
   (B) Ulf, Xavi, Victor
   (C) Victor, Tom
   (D) Sam, Victor
   (E) Victor, Tom, Sam

Supermodels

Exactly six supermodels – Anka, Bianca, Casablanca, Dominique, Ennui, and Fromage – will walk the runway at a fashion show. There are seven time slots in the fashion show. Each supermodel will walk exactly once, and there will be one intermission. The supermodels will walk according to the following conditions:

Casablanca must walk immediately before Dominique.
The intermission will be scheduled sometime after Casablanca walks.
There are exactly two supermodels scheduled to walk in between the time Anka or Ennui walks.

1. If Ennui walks fifth, then which of the following will be the fourth timeslot?
   (A) Intermission
   (B) Dominique
   (C) Casablanca
   (D) Anka
   (E) Fromage

2. If Anka walks first, then Casablanca could walk:
   (A) Second
   (B) Third
   (C) Fourth
   (D) Sixth
   (E) Seventh

3. If Bianca and Fromage are scheduled as far apart from each other as possible, then which of the following is an acceptable list of the supermodels who can walk in timeslots one through three?
   (A) Anka, Bianca, Casablanca
   (B) Fromage, Anka, Intermission
   (C) Bianca, Casablanca, Dominique
   (D) Casablanca, Dominique, Fromage
   (E) Bianca, Anka, Casablanca

4. The Intermission cannot be scheduled in which of the following slots?
   (A) Second
   (B) Third
   (C) Fourth
   (D) Fifth
   (E) Sixth

5. What is the latest timeslot in which Casablanca can walk?
   (A) Third
   (B) Fourth
   (C) Fifth
   (D) Sixth
   (E) Seventh

6. The maximum number of timeslots that can separate Fromage from the Intermission is:
   (A) 1
   (B) 2
   (C) 3
   (D) 4
   (E) 5

7. Dominique has only one possible slot to walk if Anka walks:
   (A) First
   (B) Third
   (C) Fourth
   (D) Fifth
   (E) Sixth

Suspects in a lineup

Six suspects to a crime – Guilty Gabe, No-Good Ned, Rough Ralph, Shady, Two-bit, and Wylie Willie – are brought into the police department for a Witness Identification lineup. There are five spots in the lineup numbered 1 through 5 from left to right. Five of the six suspects will be placed in the lineup according to the following conditions:

The Witness will ID the suspect in spot 2 or spot 4 as the criminal from the robbery.

If he is in the lineup, Guilty Gabe can only be in spots 2, 3, or 4.

No-Good Ned and Rough Ralph will be next to each other in the lineup.

If Shady is in the lineup, then Wylie Willie will be in spot 4 or 5.

1. If Shady is in spot 5, which of the following must be false?
   (A) No-Good Ned is in spot 1.
   (B) Guilty Gabe is in spot 2.
   (C) Guilty Gabe is in spot 3.
   (D) Guilty Gabe is not in the lineup.
   (E) Two-Bit is not in the lineup.

2. Which of the following could be a partial list of the suspects in spots 1 through 3, respectively?
   (A) Two-Bit, Guilty Gabe, Shady
   (B) Wylie, Guilty Gabe, Shady
   (C) No-Good Ned, Rough Ralph, Shady
   (D) No-Good Ned, Guilty Gabe, Two-Bit
   (E) Wylie, Two-Bit, No-Good Ned

3. If the Witness identifies Two-Bit as the criminal, then which of the following is a complete list of suspects that could be in spot 2?
   (A) Guilty Gabe, No-Good Ned, Rough Ralph, Shady, Wylie Willie
   (B) Guilty Gabe, No-Good Ned, Rough Ralph, Two-bit
   (C) Guilty Gabe, Shady, Wylie Willie, Two-bit
   (D) No-Good Ned, Rough Ralph, Two-bit
   (E) Rough Ralph, Two-bit

4. Which of the following could be true?
   (A) Wylie Willie is not in the lineup.
   (B) No-Good Ned is not in the lineup.
   (C) Shady is not in the lineup.
   (D) Two-bit is always in the lineup.
   (E) Guilty Gabe is always in the lineup.

5. Suppose that the Witness identifies No-Good Ned as the criminal. Which of the following could be the lineup order from 1 to 5?
   (A) Two-Bit, Shady, Rough Ralph, No-Good Ned, Guilty Gabe
   (B) Two-Bit, Guilty Gabe, No-Good Ned, Rough Ralph, Wylie Willie
   (C) Shady, No-Good Ned, Guilty Gabe, Rough Ralph, Wylie Willie
   (D) Wylie Willie, Shady, Rough Ralph, No-Good Ned, Two-Bit
   (E) Wylie Willie, Guilty Gabe, Rough Ralph, No-Good Ned, Two-Bit

Geography Lesson

A school geography teacher has dedicated six months – September, October, November, December, January, and February – to teach about six regions of the world – Africa, Asia, Europe, North America, South America, and the United States. Each region will be taught for exactly one month according to the following conditions:

Asia is taught in either September or October.
Europe is taught exactly two months before the United States.
North America is not taught in November.
If Africa is taught in November then South America is taught in January or February.

1. Which an acceptable order for teaching the regions from September to February?
   (A) Asia, Africa, North America, Europe, South America, United States
   (B) Africa, Asia, Europe, North America, South America, United States
   (C) Africa, Asia, Europe, South America, United States, North America
   (D) Asia, South America, Africa, Europe, North America, United States
   (E) North America, South America, Europe, Asia, United States, Africa

2. If South America is taught in September, which of the following could be false?
   (A) United States is taught in a later month than Asia.
   (B) Europe is taught in November.
   (C) United States is taught in January.
   (D) Asia is taught in October.
   (E) United States is taught in a later month than Africa.

3. If Africa is taught in November, how many possible months could Europe be taught?
   (A) 1
   (B) 2
   (C) 3
   (D) 4
   (E) 5

4. If United States is taught in November, each of the following could be true EXCEPT:
   (A) United States is taught before North America.
   (B) Asia is taught before North America.
   (C) South America is taught before North America.
   (D) South America is taught before United States.
   (E) South America is taught before Africa.

5. If North America is taught during the month in between the months that Europe and United States are taught, which of the following is a complete and accurate list of the months in which United States could be taught:
   (A) January, February
   (B) January, November
   (C) November, December, January
   (D) November, December, January, February
   (E) November, January, February

Wine and Cheese Pairing

A trendy new wine bar offers fine wine flights paired with artisan cheeses. The bar offers the following red wines – Bordeaux from France, Malbec from Argentina, Shiraz from South Africa, and Dolcetto d'Alba from Italy. The bar also offers the following white wines – Piñot Gris from France, Riesling from Germany, and Chardonnay from Napa Valley. Each wine flight consists of three different wines. Customers can order a Red Flight (only red wines), a White Flight (only white wines), or a Mixed Flight (both red and white wines).

Each wine flight will be paired with a flight of three different cheeses. The bar offers the following cheeses – Ambert from France, Emmental from Switzerland, Idiazabal from Spain, Oaxaca from Mexico, and Ubriaco from Italy. Further, the wines and cheeses must be paired according to the following restrictions:

Dolcetto pairs with Ubriaco.
Malbec pairs with either Oaxaca or Idiazabal.
Pinot Gris, Bordeaux, and Shiraz all pair with Ambert.

1. Which of the following could be the list of cheeses paired with a Red Flight?
   (A) Ambert, Emmental, Idiazabal
   (B) Ambert, Emmental, Oaxaca
   (C) Ambert, Oaxaca, Idiazabal
   (D) Ambert, Idiazabal, Ubriaco
   (E) Idiazabal, Oaxaca, Ubriaco

2. If Pinot Gris and Emmental are served then which of the following must be true?
   (A) The flight consists of two red wines and one white wine.
   (B) The flight consists of two white wines and one red wine.
   (C) The flight is Red or Mixed.
   (D) The flight is White or Mixed.
   (E) The flight is Red or White.

3. If a flight is ordered that consists only of European (Italy, France, and Germany) wines and cheeses, which of the following cannot be true?
   (A) Ambert and Emmental are served.
   (B) Ambert and Ubriaco are served.
   (C) Emmental and Ubriaco are served.
   (D) Emmental and Idiazabal are served.
   (E) Ubriaco and Idiazabal are served.

4. If Oaxaca and Idiazabal are served, then which of the following cannot be the list of wines in the flight?
   (A) Malbec, Riesling, Pinot Gris
   (B) Malbec, Riesling, Chardonnay
   (C) Malbec, Bordeaux, Dolcetto
   (D) Bordeaux, Riesling, Chardonnay
   (E) Malbec, Riesling, Shiraz

5. If Ambert is not served, then which of the following could be true?
   (A) The flight is Red.
   (B) The flight is White.
   (C) The flight is Mixed.
   (D) The flight is Red or Mixed.
   (E) The flight is White or Mixed.

6. Which of the following is an acceptable Wine and Cheese Flight?
   (A) Wines: Dolcetto, Malbec, Shiraz
   Cheeses: Ubriaco, Oaxaca, Emmental
   (B) Wines: Dolcetto, Malbec, Bordeaux
   Cheeses: Oaxaca, Idiazabal, Ambert
   (C) Wines: Dolcetto, Bordeaux, Shiraz
   Cheeses: Ubriaco, Ambert, Ambert
   (D) Wines: Pinot Gris, Riesling, Shiraz
   Cheeses: Ambert, Emmental, Ubriaco
   (E) Wines: Pinot Gris, Riesling, Chardonnay
   Cheeses: Ambert, Emmental, Idiazabal

Russian Composers

As part of a special Night of Tribute to Russian Composers, the London Philharmonic Orchestra will perform the works of several Russian composers. The Orchestra will play 1 piece by Obukhov, 2 pieces by Prokofiev, 2 pieces by Rachmaninoff, 1 piece by Stravinsky, 2 pieces by Tchaikovsky, and 1 piece by Ustvolskaya. The pieces will be played according to the following rules:

The first Prokofiev piece will be played before the first Rachmaninoff piece.
The first Rachmaninoff piece will be played before the second Prokofiev piece.
Stravinsky will be played before both Tchaikovsky pieces.
Ustvolskaya will be played before Stravinsky and both Rachmaninoff pieces.

1. Which one of the following could be a complete and accurate list of the pieces played in order?
   (A) Prokofiev, Stravinsky, Ustvolskaya, Prokofiev, Rachmaninoff, Tchaikovsky, Rachmaninoff, Tchaikovsky, Obukhov
   (B) Prokofiev, Ustvolskaya, Obukhov, Rachmaninoff, Prokofiev, Tchaikovsky, Stravinsky, Rachmaninoff, Tchaikovsky
   (C) Stravinsky, Prokofiev, Obukhov, Ustvolskaya, Tchaikovsky, Tchaikovsky, Rachmaninoff, Rachmaninoff, Prokofiev
   (D) Prokofiev, Ustvolskaya, Rachmaninoff, Stravinsky, Tchaikovsky, Tchaikovsky, Prokofiev, Rachmaninoff, Obukhov
   (E) Obukhov, Prokofiev, Ustvolskaya, Stravinsky, Tchaikovsky, Tchaikovsky, Prokofiev, Rachmaninoff, Rachmaninoff

2. Which of the following composers cannot be played last?
   (A) Obukhov
   (B) Prokofiev
   (C) Rachmaninoff
   (D) Stravinsky
   (E) Tchaikovsky

3. If Ustvolskaya is played second, which of the following could not be played third?
   (A) Obukhov
   (B) Prokofiev
   (C) Rachmaninoff
   (D) Stravinsky
   (E) Tchaikovsky

4. If the second Rachmaninoff piece is played directly in between the two Tchaikovsky pieces, then which of the following must be false?
   (A) Tchaikovsky is played fourth.
   (B) Tchaikovsky is played fifth.
   (C) Tchaikovsky is played sixth.
   (D) Tchaikovsky is played seventh.
   (E) Tchaikovsky is played eighth.

5. If Prokofiev is played first and ninth, then which of the following could be true?
   (A) Obukhov is played third and Ustvolskaya is played fourth.
   (B) Ustvolskaya is played third and Obukhov is played fourth.
   (C) Rachmaninoff is played seventh and eighth.
   (D) Stravinsky is played seventh and Tchaikovsky is played eighth.
   (E) Stravinsky is played seventh and Obukhov is played eighth.

6. Which of the following could not be a partial list of the first four composers played?
   (A) Ustvolskaya, Stravinsky, Tchaikovsky, Prokofiev
   (B) Prokofiev, Ustvolskaya, Rachmaninoff, Stravinsky
   (C) Obukhov, Prokofiev, Stravinsky, Tchaikovsky
   (D) Obukhov, Ustvolskaya, Prokofiev, Rachmaninoff
   (E) Obukhov, Prokofiev, Ustvolskaya, Rachmaninoff

Desert Safari

Seven friends – Aisha, Bilal, Chamis, Diyaa, Esam, Farhan, and Grace – are driving a rugged SUV through the peaks and valleys of the sand dunes outside of Dubai. The SUV has seven seats and three rows – two seats in the first row, two seats in the middle row, and three seats in the back row. All the seats are window seats except the back row middle seat.

The first row will consist of the Driver on the left and the Navigator on the right. The two middle seats are directly behind the two first row seats. The back row window seats are directly behind the middle row seats, and the back row middle seat is behind no one. People with motion sickness cannot sit in the back row. The friends will be seated in the SUV according to the following rules:

Aisha is either Driver or Navigator.
Bilal and Diyaa cannot sit in the same row.
Grace gets bad motion sickness.
Farhan sits directly behind Grace.

1. If Chamis is the Navigator and Grace sits directly behind her, then which of the following must be true?
   (A) Bilal is in the back row.
   (B) Diyaa is in the back row.
   (C) Diyaa is in the middle row.
   (D) Esam is in the middle row.
   (E) Esam is in the back row.

2. The back middle seat cannot be occupied by:
   (A) Bilal
   (B) Farhan
   (C) Diyaa
   (D) Chamis
   (E) Esam

3. If Aisha is the Driver and Esam sits directly behind her, then each of the following could be true EXCEPT:
   (A) Bilal is the Navigator.
   (B) Diyaa is the Navigator.
   (C) Grace is the Navigator.
   (D) Chamis in the back row.
   (E) Farhan is in the back row.

4. If Grace is not seated next to Bilal or Diyaa, then which of the following must be false?
   (A) Grace sits in the middle row.
   (B) Grace is the driver.
   (C) Bilal is the driver.
   (D) Chamis is the driver.
   (E) Chamis sits in the back row.

5. Which of the following is an acceptable partial list of people who can occupy the back row from left to right?
   (A) Chamis, Bilal, Diyaa
   (B) Chamis, Esam, Grace
   (C) Esam, Farhan, Chamis
   (D) Chamis, Esam, Farhan
   (E) Bilal, Chamis, Aisha

Pirate Plank Walk

A feisty band of Pirates has unfortunately been captured by an enemy ship and taken onboard as prisoners. There are six Pirates that have been taken prisoner – BlueBeard, Calico Jack, Diabolito, Eric the Red, GreenBeard, and Long John. The enemies will make the prisoner Pirates walk the Plank.

The enemy captain has lined up the prisoner Pirates on deck standing shoulder to shoulder in a horizontal line so that he can make a threatening speech. He will then force the prisoner Pirates to walk the plank one by one, either from left to right, or from right to left, in accordance with the following rules:

Bluebeard is standing immediately to the left of GreenBeard.
Calico Jack is standing to the left of everyone.
Diabolito and Calico Jack are not standing next to each other.
The fifth pirate to walk the plank will be granted Parley at the last minute, and be spared by the enemies.

1. If Long John walks the plank fourth, then which one of the following must be false?
   (A) BlueBeard walks the plank third.
   (B) GreenBeard walks the plank third.
   (C) Diabolito walks the plank second.
   (D) Eric the Red walks the plank third.
   (E) Eric the Red walks the plank second.

2. If Eric the Red is standing immediately left of Long John, then which one of the following could be true?
   (A) Eric the Red walks third and the order is from left to right.
   (B) Long John walks third and the order is from right to left.
   (C) Diabolito is granted Parley and the order is from left to right.
   (D) GreenBeard is granted Parley and the order is from right to left.
   (E) GreenBeard is granted Parley and the order is from left to right.

3. If BlueBeard is standing next to the Pirate that is granted Parley, then which of the following is a complete list of the Pirates who could be granted Parley?
   (A) Diabolito, GreenBeard, Eric the Red, Long John
   (B) GreenBeard, Eric the Red, Long John
   (C) Diabolito, Eric the Red, Long John
   (D) Eric the Red, Long John
   (E) GreenBeard

4. Each of the following could be true whether the pirates walk left to right or right to left EXCEPT:
   (A) Long John walks the plank second.
   (B) Long John walks the plank third.
   (C) BlueBeard walks the plank second.
   (D) GreenBeard walks the plank second.
   (E) GreenBeard walks the plank third.

5. Which one of the following pirates can never walk the plank first?
   (A) BlueBeard
   (B) Eric the Red
   (C) Diabolito
   (D) GreenBeard
   (E) Long John

6. Which one of the following is an acceptable partial list of the first three pirates to walk the plank?
   (A) First: Calico Jack; Second: Eric the Red; Third: GreenBeard
   (B) First: Eric the Red; Second: Long John; Third: GreenBeard
   (C) First: Eric the Red; Second: BlueBeard; Third: GreenBeard
   (D) First: Diabolito; Second: Eric the Red; Third: GreenBeard
   (E) First: Long John; Second: Diabolito; Third: BlueBeard

Senate Committees

A government is forming three committees – Budget, Environment, and Security. The committees will be comprised of ten senators – three from the north, three from the west, two from the east, and two from the south. The committees will be organized according to the following conditions:

Each committee has at least three senators.
The two southern senators serve in different committees.
The Budget committee has at least one northern senator.
If the Environment committee has a senator from the west, both eastern senators serve in the Budget committee.
If a western senator serves in the Security committee, then all senators from the west serve in the Security committee.

1. If all the senators from the north serve in the Budget committee each of the following must be true EXCEPT:
   (A) A western senator serves in the Security committee.
   (B) An eastern senator serves in the Environment committee.
   (C) A southern senator serves in the Environment committee.
   (D) Exactly three senators serve in the Environment committee.
   (E) Exactly three senators serve in the Security committee.

2. If exactly two western senators serve on the Environment committee, for how many of the senators can their committees be determined?
   (A) 6
   (B) 7
   (C) 8
   (D) 9
   (E) 10

3. If no southern senator serves in the Environment committee, each of the following could be true EXCEPT:
   (A) A northern senator serves in the Environment committee.
   (B) A northern senator serves in the Security committee.
   (C) A western senator serves in the Security committee.
   (D) An eastern senator serves in the Security committee.
   (E) A western senator serves in the Environment committee.

4. If exactly four senators serve in the Security committee, which must be false?
   (A) A western senator serves in the Environment committee.
   (B) A northern senator serves in the Environment committee.
   (C) An eastern senator serves in the Environment committee.
   (D) A southern senator serves in the Environment committee.
   (E) A southern senator serves in the Budget committee.

5. Which of the following is an acceptable assignment of senators to committees?
   (A) Budget: north, north, west
   Environment: west, west, south
   Security: south, east, east, north
   (B) Budget: north, east, east, south
   Environment: south, west, north
   Security: west, west, north
   (C) Budget: north, north, east, east
   Environment: north, south, east, south
   Security: west, west, west
   (D) Budget: north, east, east
   Environment: west, west, west, south
   Security: north, south, north
   (E) Budget: south, east, east
   Environment: north, north, south
   Security: west, west, west, north

Symposium Speakers

Nine speakers – Professor Ickles, Reverend Johnson, Professor Keats, Professor Lemke, Reverend Mitchell, Reverend Nielsen, Professor O'Hara, Professor Plum, and Reverend Quintess – are scheduled to speak at a university symposium about Science and Religion. The speakers will speak one at a time pursuant to the following conditions:

Professor Plum speaks before Professor O'Hara and Reverend Mitchell.
Reverend Mitchell speaks after Reverend Nielsen.
Professor O'Hara speaks before Reverend Johnson and Professor Keats.
Reverend Johnson speaks before Reverend Quintess, but after Professor Ickles.

1. Which of the following is an acceptable order of speakers?
   (A) Lemke, Plum, Nielsen, O'Hara, Keats, Johnson, Ickles, Quintess, Mitchell
   (B) Plum, Nielsen, Lemke, Mitchell, Ickles, Johnson, Keats, O'Hara, Quintess
   (C) Plum, Keats, O'Hara, Nielsen, Lemke, Ickles, Johnson, Quintess, Mitchell
   (D) Nielsen, Lemke, Plum, Mitchell, O'Hara, Keats, Ickles, Quintess, Johnson
   (E) Plum, Lemke, Nielsen, O'Hara, Mitchell, Keats, Ickles, Johnson, Quintess

2. If Keats is the fifth speaker, which of the following cannot be true?
   (A) Two professors and two reverends speak before Keats.
   (B) Three professors and one reverend speak before Keats.
   (C) Three reverends and one professor speak before Keats.
   (D) Four professors speak before Keats.
   (E) Two professors and two reverends speak after Keats.

3. Which of the following is a complete list of speakers who could speak last?
   (A) Johnson, Keats, Lemke, Mitchell, Nielsen, Quintess
   (B) Keats, Lemke, Mitchell, Quintess
   (C) Lemke, Mitchell, Nielsen
   (D) Quintess, Mitchell
   (E) Quintess, Mitchell, Nielsen

4. If Lemke speaks first, then which of the following must be true?
   (A) Johnson can speak no earlier than fifth.
   (B) Ickles can speak no earlier than fourth.
   (C) Quintess can speak no earlier than sixth.
   (D) The final speaker is a reverend.
   (E) The final speaker is a professor.

5. If Mitchell is the second reverend to speak, and speaks fourth, how many different speakers can speak third?
   (A) 2
   (B) 3
   (C) 4
   (D) 5
   (E) 6

Games

PART 2

Parasailing

In Jamaica, five guys – Amani, Bowie, Clifton, Demonde, and Eldrick – and four girls – Frida, Galilea, Haven, and Irie – are going parasailing. They will take their boat from Reggae Beach around the coast to Sapphire Island for a picnic lunch, and then make a return journey back to Reggae Beach in the evening.

There are seven seats on the boat and two seats on the parasail. Each of the nine friends will be one of five possible positions on each journey. At the very front of the boat, one person will pilot the boat. Behind the pilot there are three people seated on the front bench. Behind the front bench, there are two people seated on the back bench. Behind the back bench there will be one person who acts as the line keeper, maintaining the line that connects the boat to the parasail. Behind the line keeper will be two lucky people who get to sit in the parasail as the boat travels through the water. The friends will be arranged on the boat according to the following rules:

Amani will either pilot the boat or keep the line.
Either Clifton or Irie must be on the parasail, but never together.
Frida and Haven always sit next to each other.
Eldrick sits in a position somewhere in front of Frida.
No one sits in the same positions on both the outbound and return journeys.

1. Which of the following is a complete list of people who could pilot the boat?
   (A) Amani
   (B) Amani, Eldrick
   (C) Amani, Eldrick, Clifton
   (D) Amani, Clifton
   (E) Amani, Clifton, Eldrick, Irie

2. If Bowie and Galilea sit next to each other on the outbound journey, then which of the following cannot be true of the return journey?
   (A) Amani is the pilot.
   (B) Clifton is the line keeper.
   (C) Irie is the line keeper.
   (D) Galilea sits in the parasail.
   (E) Demonde sits in the parasail.

3. If Bowie sits in the parasail on the outbound journey and is the line keeper on the return journey, then which of the following cannot be true?
   (A) Amani is line keeper on the outbound journey.
   (B) Eldrick is on the front bench on the return journey.
   (C) Clifton is on the back bench on the return journey.
   (D) Clifton is on the front bench on the return journey.
   (E) Demonde is on the back bench on the outbound journey.

4. If the front bench is seated with Eldrick, Bowie, and Demonde on the outbound journey, then which of the following must be true?
   (A) The keeper is male on the outbound journey.
   (B) The keeper is female on the outbound journey.
   (C) The keeper is male on the return journey.
   (D) The keeper is female on the return journey.
   (E) The keeper is the same gender on both journeys.

5. If the parasail is seated with females on the outbound journey, then which of the following could be true?
   (A) The parasail is seated with females on the return journey.
   (B) No females sit on the benches on the return journey.
   (C) All the females sit on the benches on the return journey.
   (D) All the males sit on the benches on the return journey.
   (E) The pilot and keeper are females on the return journey.

6. Which of the following cannot be true?
   (A) Amani sits in a position behind Frida.
   (B) Amani sits in a position behind Clifton.
   (C) Irie sits in a position behind Clifton.
   (D) Eldrick sits in a position behind Amani.
   (E) Eldrick sits in a position behind Clifton.

Singing Competition

A music competition is being held in New York. The competition has three girls – Alice, Beth, and Claire – and three boys – Rick, Sam, and Tom. All six competitors will sing on Wednesday and then again on Friday. The competitors will each have six minutes to sing and then four minutes of feedback in the following p.m. timeslots: 8:00, 8:10, 8:20, 8:30, 8:40, and 8:50. The competition singing schedule will be established according to the following conditions:

No two boys and no two girls sing consecutively on Wednesday.
Beth is the last girl to perform each day.
Sam sings in the same timeslot on Wednesday and Friday.
Each day, Tom sings sometime before Claire.

1. If Tom sings at 8:10 on Wednesday, which of the following could be true?
   (A) Rick sings at 8:40 on Friday.
   (B) Sam sings at 8:40 on Friday.
   (C) Claire sings at 8:00 on Wednesday.
   (D) Alice sings at 8:20 on Wednesday.
   (E) Beth sings at 8:30 on Wednesday.

2. If Rick is the first boy to sing in the competition, how many of the timeslots can be determined?
   (A) 3
   (B) 4
   (C) 5
   (D) 7
   (E) 10

3. If the schedule was set so the maximum amount of time was passed between Alice's first and last performance, which of the following could be false?
   (A) Beth sings before Rick on Wednesday.
   (B) Tom sings before Rick on Friday.
   (C) Alice sings after Sam on Friday.
   (D) Sam sings before Rick on Wednesday.
   (E) Tom sings before Sam on Friday.

4. What is the total number of different timeslots on which Beth could sing over the two days?
   (A) 2
   (B) 3
   (C) 4
   (D) 5
   (E) 6

5. If on Friday, Alice sang in the timeslot immediately after Sam's timeslot, which of the following must be true?
   (A) The latest Alice can sing on Wednesday is 8:50.
   (B) The earliest Alice can sing on Wednesday is 8:20.
   (C) The earliest Tom can sing on Wednesday is 8:10.
   (D) The latest Sam can sing on Wednesday is 8:30.
   (E) The earliest Claire can sing on Wednesday is 8:20.

6. Which is an acceptable competition schedule:

		8:00	8:10	8:20	8:30	8:40	8:50
(A)	W:	Alice	Tom	Sam	Claire	Beth	Rick
	F:	Tom	Claire	Sam	Alice	Rick	Beth
(B)	W:	Alice	Tom	Claire	Rick	Beth	Sam
	F:	Tom	Claire	Beth	Alice	Rick	Sam
(C)	W:	Tom	Claire	Sam	Alice	Rick	Beth
	F:	Tom	Alice	Claire	Sam	Beth	Rick
(D)	W:	Tom	Claire	Rick	Alice	Sam	Beth
	F:	Tom	Alice	Claire	Beth	Sam	Rick
(E)	W:	Alice	Tom	Sam	Claire	Beth	Rick
	F:	Alice	Claire	Sam	Tom	Rick	Beth

Heroes vs. Villains

A battle is being fought between eight fighters. Four fighters are heroes – Adam, Bill, Carl, and David – and four fighters are villains – Rick, Sam, Tom, and Ulf. Each hero fights against exactly one villain. Each villain fights against exactly one hero. The following weapons are available to the fighters: 3 rifles, 2 handguns, 2 knives, and 1 grenade. Each fighter uses exactly one weapon in the battle. The battle is fought according to the following conditions:

Carl has a rifle and fights Tom.
A villain has the grenade.
Sam does not fight Adam.
Exactly one villain has a rifle.
If David fights Ulf then Rick has a knife.
A hero has a different weapon than the villain he is fighting.

1. If Bill has a rifle and fights Ulf, which of the following could be true?
   (A) Adam and Rick have knives.
   (B) Carl and Ulf have rifles.
   (C) David and Sam have handguns.
   (D) Adam and Sam have handguns.
   (E) Rick and Sam have knives.

2. If the villains who fight Adam and Bill have handguns, which must be true?
   (A) Adam has a knife.
   (B) Ulf has a handgun.
   (C) Sam has a handgun.
   (D) Bill has a knife.
   (E) Rick has a rifle.

3. If the heroes have all the knives, how many hero-villain pairing combinations are possible:
   (A) 3
   (B) 4
   (C) 5
   (D) 6
   (E) 10

4. If Adam has a knife, each of the following could be true, EXCEPT:
   (A) Bill fights Rick.
   (B) Tom uses a grenade.
   (C) Ulf fights David.
   (D) Sam uses a rifle.
   (E) David fights Sam.

5. If Bill fights Sam, which of the following yields only one possible set of hero-villain pairings:
   (A) Bill has a knife.
   (B) Tom has a handgun.
   (C) Rick has a knife.
   (D) David has a handgun.
   (E) Adam has a knife.

Grand Jury Indictments

As the result of numerous financial scandals at a notorious Wall Street Financial Services Firm, a Grand Jury will hear evidence presented by the U.S. Attorney against four leading executives at the Firm. The four executives are the Chief Executive Officer (CEO), Chief Financial Officer (CFO), Chief Operating Officer (COO), and General Counsel (GC). After deliberating on the evidence, the Grand Jury must decide whether to indict each executive on three charges – Accounting Fraud, Insider Trading, and/or Stock Manipulation – pursuant to the following conditions:

The Grand Jury will indict exactly two executives for each criminal charge.

Each executive will be indicted on at least one charge.

The Chief Executive Officer and Chief Financial Officer are indicted for Stock Manipulation and Accounting Fraud, respectively, either together or not at all.

If the Grand Jury indicts the General Counsel for Insider Trading, then the Grand Jury will indict the Chief Operating Officer for Accounting Fraud.

The Grand Jury will not indict the Chief Executive Officer and the General Counsel on the same charges.

1. If the Grand Jury indicts the General Counsel and the Chief Operating Officer on Accounting Fraud, then which one of the following must be true?
   (A) The General Counsel is indicted for Stock Manipulation.
   (B) The Chief Financial Officer is indicted for Stock Manipulation.
   (C) The Chief Operating Officer is indicted for Stock Manipulation.
   (D) The Chief Financial Officer is indicted for Insider Trading.
   (E) The Chief Executive Officer is indicted for Insider Trading.

2. If the Chief Operating Officer and the General Counsel are both indicted for Stock Manipulation, then which of the following must be false?
   (A) The Chief Financial Officer is indicted on exactly one charge.
   (B) The Chief Financial Officer is indicted on exactly two charges.
   (C) The Chief Operating Officer is indicted on exactly two charges.
   (D) The Chief Executive Officer is indicted on exactly two charges.
   (E) The General Counsel is indicted on exactly two charges.

3. If the Grand Jury indicts the Chief Financial Officer on all three charges, then which of the following could be true?
   (A) The Chief Executive Officer is indicted on Accounting Fraud.
   (B) The Chief Operating Officer is indicted on Accounting Fraud.
   (C) The General Counsel is indicted on Stock Manipulation.
   (D) The Chief Operating Officer is indicted on Stock Manipulation.
   (E) The Chief Executive Officer is indicted on Insider Trading.

4. If the Grand Jury indicts the same two executives on charges of Accounting fraud and Insider Trading, those two executives could be:
   (A) The Chief Executive Officer and the General Counsel
   (B) The Chief Financial Officer and the General Counsel
   (C) The Chief Financial Officer and the Chief Operating Officer
   (D) The Chief Executive Officer and the Chief Operating Officer
   (E) The General Counsel and Chief Operating Officer

5. Each of the following could be true EXCEPT:
   (A) The Chief Financial Officer is indicted on only one charge.
   (B) The General Counsel is indicted on exactly two charges.
   (C) The Chief Operating Officer is indicted on all three charges.
   (D) The Chief Operating Officer and the Chief Financial Officer are both indicted on exactly two charges.
   (E) The Chief Operating Officer and the Chief Financial Officer are both indicted on exactly one charge each.

6. If the rule that General Counsel and Chief Executive Officer may not be indicted on the same charges is removed, which of the following must be false?
   (A) General Counsel and Chief Executive Officer are both indicted for Insider Trading and Stock manipulation.
   (B) General Counsel and Chief Executive Officer are both indicted for Accounting Fraud, only.
   (C) General Counsel and Chief Executive Officer are both indicted for Accounting Fraud and Stock Manipulation.
   (D) Chief Executive Officer is indicted for Accounting Fraud and the General Counsel is indicted for Stock Manipulation.
   (E) Chief Executive Officer is indicted for Insider Trading and General Counsel is indicted for Accounting Fraud.

Zoo Budget Cuts

Due to budget cuts a zoo must relocate six of its animals. Identified for relocation were four reptiles - an iguana, a snake, and two crocodiles; five big cats – three tigers and two lions; and two elephants. The animals will be chosen according to the following rules:

No more than four big cats are chosen.
If the iguana is chosen, exactly two tigers are chosen.
If an elephant is chosen, exactly two reptiles are chosen.
If the snake is chosen no crocodiles are chosen.
If both elephants are chosen then the snake must be chosen.

1. Which is a possible list of the animals that are chose for relocation together?
   (A) Two tigers, a snake, a crocodile, a lion, an elephant
   (B) Two tigers, an elephant, two crocodiles, a lion
   (C) Three tigers, an elephant, a snake, and a lion
   (D) Three tigers, two lions, two crocodiles
   (E) Two tigers, an iguana, two elephants, a crocodile

2. If all three tigers are chosen, each of the following must be true EXCEPT:
   (A) The iguana is not chosen.
   (B) The snake is not chosen.
   (C) Both crocodiles are chosen.
   (D) One lion is chosen.
   (E) Both elephants cannot be chosen.

3. If exactly two reptiles are chosen, which must be true?
   (A) A crocodile is chosen.
   (B) If the iguana is chosen a lion is chosen.
   (C) If the snake is chosen a lion is chosen.
   (D) If only one crocodile is chosen exactly two tigers are chosen.
   (E) If both crocodiles are chosen only one lion is chosen.

4. If only one crocodile and one lion are selected, how many possible ways can the animals be chosen?
   (A) 1
   (B) 2
   (C) 3
   (D) 4
   (E) 5

5. Suppose that the rule that "If the snake is chosen no crocodiles are chosen" is removed and replaced with the rule that "Both lions must be selected," with all other rules remaining the same. How many possible ways can the animals be chosen?
   (A) 2
   (B) 4
   (C) 5
   (D) 7
   (E) 8

International Ambassadors

An international ambassador from Australia, Brazil, China, England, India, Nigeria, Russia, and the United States are meeting to sit down and discuss world peace. The ambassadors are to be seated, evenly spaced, at a round boardroom table according the following rules:

The ambassador from United States and the ambassador from England must sit immediately next to each other.

The ambassador from China and the ambassador from Russia must sit directly across from each other.

The ambassador from Brazil sits immediately to the right of the ambassador from Nigeria.

The ambassador from Australia and the ambassador from India do not sit next to each other.

1. If the ambassador from China sits immediately to the right of the ambassador from United States, which of the following could be true?
   (A) The ambassador from Brazil sits next to the ambassador from China.
   (B) The ambassador from Nigeria sits next to the ambassador from England.
   (C) The ambassador from China sits next to the ambassador from England.
   (D) The ambassador from India sits next to the ambassador from England.
   (E) The ambassador from Russia sits next to the ambassador from England.

2. If the ambassador from India sits exactly two seats to the left of the ambassador from Australia, which must be false?
   (A) The ambassador from Brazil sits across from the ambassador from Australia.
   (B) The ambassador from Russia sits next to the ambassador from India.
   (C) The ambassador from China sits next to the ambassador from England.
   (D) The ambassador from China sits next to the ambassador from Brazil.
   (E) The ambassador from Nigeria sits next to the ambassador from Australia.

3. If the ambassador from Nigeria sits directly across from the ambassador from India, each of the following could be true EXCEPT:
   (A) The ambassador from Russia sits next to the ambassador from India.
   (B) The ambassador from Russia sits next to the ambassador from Brazil.
   (C) The ambassador from China sits next to the ambassador from India.
   (D) The ambassador from United States sits next to the ambassador from India.
   (E) The ambassador from Australia sits next to the ambassador from Brazil.

4. If ambassador from Australia sits next to the ambassador from China, which of the following must be true?
   (A) The ambassador from India sits next to the ambassador from Russia or the ambassador from China.
   (B) The ambassador from India sits next to the ambassador from Nigeria or the ambassador from United States.
   (C) The ambassador from India sits next to the ambassador from Nigeria or the ambassador from Brazil.
   (D) The ambassador from England sits next to the ambassador from Nigeria or the ambassador from Brazil.
   (E) The ambassador from England sits next to the ambassador from Russia or the ambassador from China.

5. If the condition that the ambassador from Australia and the ambassador from India do not sit next to each other is removed and replaced with the condition that the ambassador from Australia must sit directly across from the ambassador from India, how many different ambassadors could sit next to England?
   (A) 2
   (B) 3
   (C) 4
   (D) 5
   (E) 6

Cheerleading Pyramid

Ten cheerleaders – Ally, Bibi, Cici, Deedee, Ellen, Fifi, Helen, Igor, Jon, and Kiki – are practicing pyramid formations for a regional cheerleading competition. Two types of pyramids are possible. There is the 3-2-1 Pyramid with 3 people on the bottom row, 2 on the middle row, and 1 on the top row, for a total of 6 people. The other option is the 4-3-2-1 Pyramid, with 4 on the bottom, 3 on the third row, 2 on the second row, on 1 on top, utilizing all 10 people. The cheerleaders will form a pyramid according to the following constraints:

Jon and Igor must always be on the outside corners of the bottom row of any pyramid.

Bibi and Cici cannot be in the pyramid unless they are in the same row and next to each other.

Deedee and Kiki cannot be in the pyramid unless they are in the same row and next to each other.

Ally, Ellen, Helen, and Kiki cannot be in the same row.

1. If Ally is on the top of a 3-2-1 Pyramid, then which of the following must be false?
   (A) Kiki and Helen are in the pyramid but Cici is not.
   (B) Helen and Ellen are in the pyramid but Fifi is not.
   (C) Ellen and Kiki and are in the pyramid but Bibi is not.
   (D) Kiki and Deedee are in the pyramid but Fifi is not.
   (E) Bibi and Cici are in the pyramid but Deedee is not.

2. If Kiki is next to Igor in the pyramid, which of the following must be true?
   (A) Bibi is in the second row.
   (B) Ally is on top of the pyramid.
   (C) Fifi is in the second row.
   (D) Helen is in the second row.
   (E) Helen is on top of the pyramid.

3. If Helen and Fifi are in the row below Ellen, which of the following must be false?
   (A) Ellen is on top of the pyramid.
   (B) Helen and Fifi are in the middle row of a 3-2-1 pyramid.
   (C) Helen and Fifi are in the second row of a 4-3-2-1 pyramid.
   (D) Helen and Fifi are in the third row of a 4-3-2-1 pyramid.
   (E) Helen and Fifi are in the bottom row of a 4-3-2-1 pyramid.

4. In a 4-3-2-1 Pyramid, which of the following could be true?
   (A) Fifi is on the top of the pyramid.
   (B) Kiki is on the top of the pyramid.
   (C) Bibi and Cici are on the bottom row.
   (D) Kiki and Deedee are on the third row.
   (E) Kiki and Deedee are on the second row.

5. Which of the following is true of a 3-2-1 pyramid?
   (A) Either Ally or Fifi may be in the pyramid, but not both.
   (B) Either Kiki or Bibi must be in the pyramid.
   (C) Both Kiki and Bibi must be in the pyramid.
   (D) If Kiki is not in the pyramid, Ally must be in the pyramid.
   (E) If neither Kiki nor Bibi is in the pyramid, then Helen must be in the pyramid.

6. Which of the following people could be on top of a 4-3-2-1 pyramid?
   (A) Cici
   (B) Jon
   (C) Helen
   (D) Fifi
   (E) Kiki

7. If the rule that Igor and Jon must be on outside corners of the bottom row of any pyramid were removed, but other rules remained the same, which of the following would NOT be an acceptable list of cheerleaders in the bottom row of a 4-3-2-1 pyramid?
   (A) Bibi, Cici, Kiki, Deedee
   (B) Helen, Ally, Fifi, Jon
   (C) Kiki, Deedee, Fifi, Jon
   (D) Bibi, Cici, Fifi, Jon
   (E) Igor, Helen, Jon, Fifi

Real Celebutantes

Network producers are casting for a new reality show called The Real Celebutantes of Biscayne Bay, featuring wealthy ladies who get into catfights at charity fashion shows. The producers have narrowed down the potential cast members to ten ladies – Alexia, Brittany, Calista, Dionne, Ella, Fallon, Gigi, Hester, Iman, and Juliet. The producers will select five cast members for the show according to the following rules:

Each of the cast members should be in at least one feud with another cast member.
Each of the cast members should be in at least one alliance with another cast member.
Alexia is in a feud with Brittany and Calista.
Brittany and Calista are in a feud with Dionne and Ella.
Ella is in a feud with Fallon, Gigi, and Hester.
Hester is in a feud with Iman and Juliet.
Alliances are formed between any two cast members who are feuding with the same person, whether or not that person is ultimately chosen for the cast.

1. If the producers select Dionne, Ella, and Fallon for the show, then the remaining two cast members could be:
   (A) Juliet and Gigi
   (B) Juliet and Hester
   (C) Calista and Hester
   (D) Alexia and Gigi
   (E) Alexia and Iman

2. If the producers select Alexia, Calista, and Hester for the show, then the remaining two cast members could be:
   (A) Brittany and Dionne
   (B) Brittany and Ella
   (C) Dionne and Fallon
   (D) Fallon and Gigi
   (E) Gigi and Iman

3. Suppose that the producers want to select Fallon, Gigi, Hester, Iman, and Juliet. Which of the following is true of the cast selection?
   (A) The selection is unacceptable because Juliet has no alliance.
   (B) The selection is unacceptable because Iman has no feud.
   (C) The selection is unacceptable because Hester has no alliance.
   (D) The selection is unacceptable because Gigi has no alliance.
   (E) The selection is unacceptable because Fallon has no feud.

4. If Alexia is selected for the cast, then which one of the following ladies must also be in the cast?
   (A) Brittany
   (B) Calista
   (C) Dionne
   (D) Ella
   (E) Hester

5. Suppose that the producers want to cast the lady with the maximum number of potential alliances. The producers could choose which of the following as cast members?
   (A) Brittany, Calista, Ella, Fallon, Gigi, or Hester
   (B) Ella, Fallon, Gigi, or Hester
   (C) Brittany, Calista, Gigi, or Hester
   (D) Ella or Hester
   (E) Ella only

6. Which one of the following represents an acceptable list of cast members for the show?
   (A) Alexia, Ella, Dionne, Fallon, Gigi
   (B) Alexia, Brittany, Dionne, Ella, Juliet
   (C) Brittany, Calista, Dionne, Hester, Iman
   (D) Brittany, Calista, Ella, Hester, Iman
   (E) Ella, Dionne, Fallon, Gigi, Hester

7. If the producers want each cast member to have at least two feuds, which of the following is an acceptable group of cast members?
   (A) Alexia, Brittany, Calista, Dionne, Ella
   (B) Alexia, Brittany, Calista, Dionne, Hester
   (C) Alexia, Brittany, Ella, Hester, Juliet
   (D) Brittany, Calista, Ella, Hester, Iman
   (E) Ella, Fallon, Gigi, Hester, Juliet

The Puma

"The Puma" is a reality dating game show in which an attractive single woman in her thirties (the Puma) will search for true love among handsome young men in their twenties. The show is midseason so there are eight contestants left – Rico, Tate, Upton, Vaughn, Wes, Xavier, Yardley, and Zane. On this week's episode, each contestant will go on a date with the Puma and will either receive a Carnation or be eliminated.

Each contestant will go on exactly one date, and no one will receive more than one Carnation. Exactly one contestant will go on a Solo Date. At the end of the Solo Date, the Puma must either save the contestant by giving him a Carnation, or eliminate that contestant immediately. Exactly five contestants will go on a Group Date. At the end of the Group Date, three contestants will win Carnations and two contestants will be eliminated. Exactly two contestants will go on the dreaded 2-on-1 Date, where two contestants will spend an awkward night of romantic rivalry. One contestant will get the Carnation and the other will be eliminated immediately. The dates, and their outcomes, will be pre-arranged by show's producers pursuant to the following conditions:

Of Vaughn and Upton, one will be eliminated and the other will get a Carnation.
Yardley will win a Carnation on the Solo Date or the Group Date.
Yardley, Rico, and Zane are on different dates.
Wes and Xavier are on the same date.
Rico will win a Carnation unless he goes on the Solo Date.

1. If Zane is eliminated on the Group Date, which of the following must be false?
   (A) Upton gets a Carnation on the 2-on-1 Date.
   (B) Upton is eliminated on Group Date.
   (C) Upton is eliminated on the 2-on-1 Date.
   (D) Vaughn gets a Carnation on the Group Date.
   (E) Vaughn is eliminated on the 2-on-1 Date.

2. If on the Group Date, Tate gets a Carnation but Vaughn is eliminated, then which of the following contestants must be eliminated?
   (A) Wes
   (B) Xavier
   (C) Rico
   (D) Upton
   (E) Zane

3. Which one of the following represents an accurate list of contestants that could be eliminated on the Solo Date?
   (A) Rico
   (B) Rico, Zane
   (C) Rico, Yardley, Zane
   (D) Rico, Zane, Upton, Tate
   (E) Zane, Upton, Tate

4. Which one of the following could be a complete and accurate list of contestants who, together, receive Carnations on this week's episode?
   (A) Yardley, Rico, Tate, Zane
   (B) Yardley, Rico, Upton, Vaughn
   (C) Yardley, Wes, Xavier, Upton
   (D) Yardley, Wes, Xavier, Rico, Tate, Vaughn
   (E) Tate, Vaughn, Wes, Zane

5. If Rico, Wes, and Xavier are eliminated, each of the following must be false EXCEPT:
   (A) Yardley goes on the Solo Date.
   (B) Rico goes on the 2-on-1 Date.
   (C) Tate goes on the Group Date.
   (D) Tate goes on the 2-on-1 Date.
   (E) Zane goes on the Group Date.

6. If Tate gets a Carnation on the 2-on-1 Date, which of the following could be true?
   (A) Rico is eliminated on the Group Date.
   (B) Rico is eliminated on the Solo Date.
   (C) Wes and Xavier win Carnations on the Group Date.
   (D) Wes and Xavier are eliminated on the Group Date.
   (E) Zane wins a Carnation on the Solo Date.

7. Which of the following must be true?
   (A) If there are three eliminations, then Yardley is on the Solo Date.
   (B) If there are three eliminations, then Rico is on the Group Date.
   (C) If there are four eliminations, then Zane is on the Solo Date.
   (D) If there are four eliminations, then Rico is on the Solo Date.
   (E) If there are four eliminations, then Yardley is on the Group Date.

Speed Dating

Seven women - Anya, Bree, Cerise, Danielle, Emmy, Fiona, and Greta – and seven men – Tomas, Ulrich, Victor, Will, Xavier, Youssef, and Zubin – attend a speed dating event. Anya, Bree, Ulrich, Victor, and Will are regular smokers; Greta and Tomas are social smokers; and all the others are nonsmokers. Anya, Cerise, Emmy, Victor, Xavier, and Zubin are liberal; Greta and Tomas are moderates; and all others are conservative. The event planners will start arranging dates between the men and women as they arrive at the event. The dates will be arranged according to the following constraints:

Men will date women.
Nonsmokers will not date regular smokers.
Liberals and conservatives will not date each other.
No one will go on more than one date.

1. Which of the following represents an acceptable list of dates?
   (A) Anya dates Tomas, Bree dates Ulrich, Cerise dates Youssef
   (B) Anya dates Tomas, Bree dates Youssef, Cerise dates Xavier
   (C) Anya dates Will, Bree dates Ulrich, Cerise dates Tomas
   (D) Anya dates Xavier, Bree dates Will, Danielle dates Zubin
   (E) Anya dates Victor, Bree dates Ulrich, Cerise dates Xavier

2. Which of the following represents a complete and accurate list of the men that Emmy could date?
   (A) Xavier
   (B) Xavier, Zubin
   (C) Xavier, Zubin, Tomas
   (D) Zubin, Tomas
   (E) Xavier, Tomas

3. If Greta goes on a date with Youssef, then the maximum number of people who don't get a date is:
   (A) 0
   (B) 1
   (C) 2
   (D) 3
   (E) 4

4. The following is a list of people who will always get a date:
   (A) Anya, Bree, Greta, Victor, Tomas, Youssef
   (B) Anya, Greta, Tomas, Youssef
   (C) Bree, Greta, Tomas, Youssef
   (D) Cerise, Emmy, Greta, Xavier, Zubin, Tomas
   (E) Greta, Tomas

5. If all of the conservative regular smokers get a date, which of the following must be true?
   (A) Ulrich dates a moderate.
   (B) Will dates a moderate.
   (C) Bree dates a moderate.
   (D) Tomas dates a conservative.
   (E) Greta dates a conservative.

Whale Watching

Five friends—Gaston, Hermes, Iphigenie, Jacques, and Katriane— go on a whale watching boat tour to see whales off the coast of Alaska. The boat consists of a top deck that has standing room for three passengers and a bottom deck with seats for the remaining two passengers. The top deck is the preferred area to see whales but the treacherous water can lead to severe motion sickness, which must be avoided. Encountering whales can be unpredictable, so the boat will take multiple trips through the coast pursuant to the following conditions:

The boat will keep making trips to sea until each passenger has seen a whale.

No one will be on the bottom deck on two consecutive trips.

Being on the top deck more than twice in a row will lead to severe motion sickness.

It takes three times on the top deck to see a whale, and no one who has already seen a whale gets to be on the top deck.

1. If Gaston, Hermes, and Iphigenie are on top deck on the first trip, then which of the following could be the list of people on the top deck on the second trip?
   (A) Gaston, Hermes, Jacques
   (B) Gaston, Hermes, Katriane
   (C) Gaston, Iphigenie, Jacques
   (D) Hermes, Iphigenie, Katriane
   (E) Hermes, Jacques, Katriane

2. If Gaston, Hermes, and Katriane are on the top deck in the first trip, and Hermes and Iphigenie are on the top deck in the fourth trip, which of the following people cannot be on the top deck in the third trip?
   (A) Gaston
   (B) Hermes
   (C) Iphigenie
   (D) Jacques
   (E) Katriane

3. If Iphigenie and Katriane are on the bottom deck in the first trip, then which of the following must be true?
   (A) Jacques and Katriane are on the top deck in the fifth trip.
   (B) Iphigenie and Katriane are on the top deck in the fifth trip.
   (C) Gaston and Hermes are on the top deck in the third trip.
   (D) Hermes is on the top deck in the second and fourth trips.
   (E) Gaston is on the top deck in the second and third trips.

4. If Gaston, Katriane, and Jacques are on top deck in the first trip, and Katriane is on top deck on the second trip, and Iphigenie is on top deck during the third trip, then which of the following could be the group of people on top deck in the fifth trip?
   (A) Gaston, Jacques, Iphigenie
   (B) Gaston, Katriane, Jacques
   (C) Katriane, Iphigenie, Jacques
   (D) Katriane, Hermes, Iphigenie
   (E) Iphigenie, Jacques, Hermes

5. What is the minimum and maximum number of trips the boat will make out to sea?
   (A) Minimum Four, Maximum Five
   (B) Minimum Four, Maximum Six
   (C) Minimum Five, Maximum Five
   (D) Minimum Five, Maximum Six
   (E) Minimum Five, Maximum Seven

6. Which of the following must be true about the passengers on the top deck in the first trip?
   (A) All three of them will be on top deck on the third trip.
   (B) All three of them will be on top deck on the fourth trip.
   (C) Exactly one of them is on top deck during the third trip.
   (D) Exactly one of them is on top deck on the fourth trip.
   (E) Exactly one of them is on top deck on the fifth trip.

Bathroom Line

In an old split level building in downtown Toronto, there is a quirky Dive Bar on the upper level, and a delicious Taco Joint on the lower level. Unfortunately, the two establishments share only one bathroom. Three Taco customers – Anuradha, Bhumika, and Chaaya – and three Bar patrons – Divya, Eashwar, and Falguni – are waiting in line to use the bathroom according to the following rules:

Bhumika gets to use the bathroom before Eashwar.
Divya uses the bathroom sometime before Chaaya.
Bhumika uses the bathroom either just before or just after Falguni.
Divya and Falguni have exactly one person in between them in line.

1. If Anuradha is third in line, which of the following could be true?
   (A) Bhumika is first in line.
   (B) Falguni is first in line.
   (C) Falguni is fourth in line.
   (D) Divya is first in line.
   (E) Divya is second in line.

2. The earliest Eashwar could be in the bathroom line is:
   (A) Second
   (B) Third
   (C) Fourth
   (D) Fifth
   (E) Sixth

3. The bathroom order is completely determined if which one of the following is true?
   (A) Chaaya is fourth in line.
   (B) Divya is first in line.
   (C) Divya is fifth in line.
   (D) Falguni is first in line.
   (E) Falguni is second in line.

4. If the first and last persons in line are Bar patrons, then which of the following must be false?
   (A) Bhumika is second in line.
   (B) Chaaya is second in line.
   (C) Anuradha is second in line.
   (D) Anuradha is third in line.
   (E) Anuradha is fourth in line.

5. If the Bar patrons are all consecutive in line, then which of the following could be true?
   (A) Bhumika is third in line.
   (B) Falguni is first in line.
   (C) Falguni is fourth in line.
   (D) Chaaya is fourth in line.
   (E) Anuradha is sixth in line.

6. If the Taco customers are second, fourth, and sixth in line, then which of the following must be true?
   (A) Divya is first in line.
   (B) Bhumika is second in line.
   (C) Divya is third in line.
   (D) Chaaya is fourth in line.
   (E) Eashwar is fifth in line.

7. Which of the following is an acceptable lineup from first to last?
   (A) Eashwar, Falguni, Bhumika, Divya, Chaaya, Anuradha
   (B) Bhumika, Falguni, Chaaya, Divya, Anuradha, Eashwar
   (C) Bhumika, Anuradha, Falguni, Chaaya, Divya, Eashwar
   (D) Divya, Anuradha, Falguni, Bhumika, Eashwar, Chaaya
   (E) Divya, Anuradha, Chaaya, Falguni, Bhumika, Eashwar

Solutions

Biff's Workout

Matching GD: Match and Order Entities **Difficult = Very Easy**

This is primarily a matching game with seven entities - the body parts – and only three spots. You must determine which workouts can be allotted to each day. Although at first glance it might appear to be an ordering game, it operates more like a matching game, as there is not a strong temporal element. The only rule that requires you to "order" is the rule that Biff must run every other day. Otherwise, you will just be matching entities to spots. The initial setup for the diagram is straightforward:

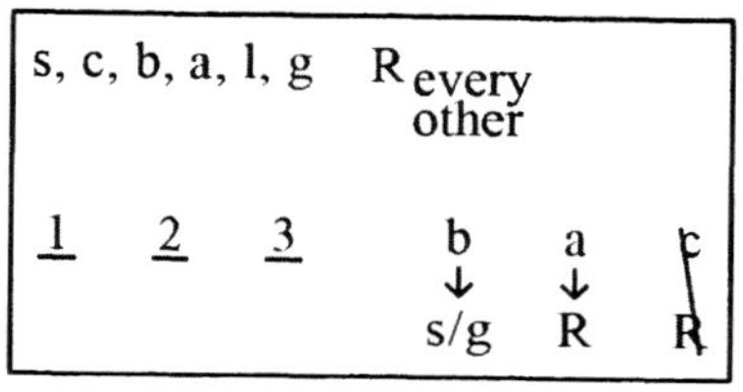

s, c, b, a, l, g R every other

1 2 3 b → s/g a → R c ↛ R

As always, scan the rules to see if any of the rules can be combined and deductions can be made. In this case, running is involved in two rules – Biff must run when he works his abs and cannot run when he works his chest. Thus, it follows that Biff cannot work out his abs and chest on the same day.

Another very obvious deduction can be made that Biff cannot work his chest, abs, or legs on the same day that he works his back, because the day he works his back he can only work his shoulders or glutes. However, this deduction does not need to be expressly drawn out as it is already well visualized as drawn.

Lastly, the fact that Biff runs every other day also conveniently limits this game. Biff can run on Day 1 *and* Day 3 or he can run on Day 2 only. Further, the deduction that he cannot run on the same day that he works his chest means that if he runs on Day 1 and Day 3 then he must work his chest on Day 2; and if he runs on Day 2 then he must work his abs on Day 2. Thus, we will always know one of the entities on Day 2 (either abs or chest) once we know which day(s) he runs and vice versa. Because Biff runs every other day, abs and chest must be worked out on consecutive days. Thus, the back can never be worked out on Day 2. As you will see, these deductions will significantly limit the game and make the game very easy.

Master Diagram

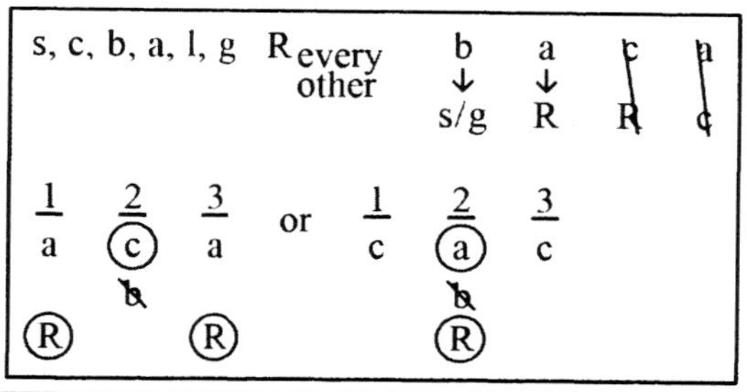

s, c, b, a, l, g R every other b → s/g a → R c ↛ R a ↛ c

1: a, (R) 2: (c), not b 3: a, (R) or 1: c 2: (a), not b, (R) 3: c

1. **(E) Biff works his back on day 2.**

If you quickly scan the answer choices, you will see that answer choice (E) is already incorporated into our master diagram. This question highlights the importance of thorough diagramming, as it can be answered in a matter of seconds if you make all the proper deductions in your master diagram. If Biff runs on day 1, he must also run on day 3 and work his chest on day 2. Therefore, he cannot work his back on day 2.

2. **(C) Biff works his glutes with his chest.**

This question has Biff working his glutes on day 2. Glutes are involved in the rule which states that Biff has to work his back with either the glutes or shoulders. If Biff works his glutes on day 2, then we know that Biff must work his back and shoulders together because, per the master diagram, he can never work his back on day 2. Biff, therefore, can work his back and shoulders together on either day 1 or day 3. The following diagram results:

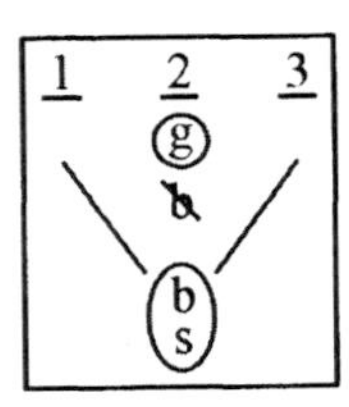

Answer choice (A) cannot be true because the back cannot be worked on day 2.

Answer choice (B) cannot be true because the back must be worked with the shoulders in this question.

Answer choice (D) cannot be true because the back must be worked with the shoulders in this question.

Answer choice (E) cannot be true because the chest is never worked with the shoulders.

3. **(A) On the day Biff works his legs, he also runs.**

This question has Biff working his chest and legs on different days. The chest, according to the rules, cannot be worked on the same day as the abs or on the same day Biff runs. The legs are not involved in any rules. Accordingly, there are three different configurations we must try if the chest and legs are on worked on different days. One where the legs are on day 1 or 3 and the chest is on day 2; one where the legs are on day 2 and the chest is on day 1 or 3; and one where the legs are on day 1/3 and the chest are on day 3/1. However, as the following diagram displays, the chest and legs cannot be on days 1 and 3, respectively:

1	2	3	1	2	3	1	2	3
(l)	(c)	(b)	(c)	(l)	(b)	(c)		(l)
(a)	~~b~~ s/g	s/g	s/g	(a)	s/g	~~b~~	~~b~~	~~b~~
(R)		(R)		(R)				

Based on the above, Biff *must* work his legs on the same day he runs (and on the same day he works his abs). Accordingly, answer choice (A) is correct. All other answer choices *could be* true, but are not necessarily true.

4. **(C) Biff works his chest on day 1 and his abs on day 3.**

If you quickly scan the answer choices, you will see that this question is about the chest and abs. Per our previous deduction, we know that one of the two has to be worked on day 2. Answer choice (C) does not have either being worked out on day 2.

5. **(E) Biff works glutes and legs on the same day.**

If Biff works his back on day 1 and runs on day 2 the second configuration in our master diagram is in play:

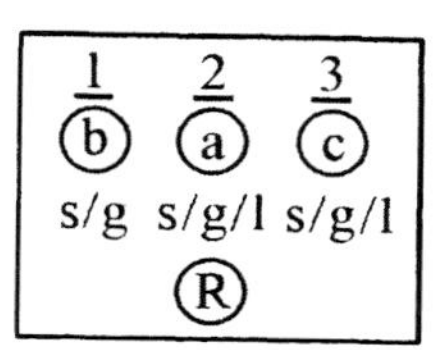

Biff cannot work his glutes and legs on the same day. Thus answer choice (E) is correct.

6. **(D) day 1: back, shoulders**

day 2: abs, glutes, run

day 3: chest, legs

Answer choice (A) is incorrect because the back is not worked out on the same day as either the shoulders or glutes.

Answer choice (B) is incorrect because Biff must run on day 3 if he runs on day 1.

Answer choice (C) is incorrect because the back is not worked out on the same day as either the shoulders or glutes.

Answer choice (E) is incorrect because Biff does not run the day he works his chest.

Scheduling Patients

Assigned Ordering GD: Order Entities **Difficulty = Very Easy**

This is a simple assigned ordering game in which you must order seven patients (entities) in seven time slots (spots). There are two positional rules and two relational rules that lend to a familiar-looking diagram:

Master Diagram

S, T, U, V, W, X, Y

9	10	11	1	2	3	4	
		Y				Y	UX or XU
~~S~~ ~~X~~	~~S~~					~~W~~	W...V

1. **(E) Walt, Xavi, Ulf, Victor, Tom, Sam, Yasser**

Answer choice (A) is incorrect because Sam cannot have the 10am time slot.

Answer choice (B) is incorrect because Yasser can only have the 11am and 4pm time slots.

Answer choice (C) is incorrect because Ulf and Xavi must be consecutive.

Answer choice (D) is incorrect because Walt is not scheduled before Victor.

2. **(B) Walt's appointment is at 1pm and Victor's appointment is at 2pm.**

This question has Tom in the 11am appointment. Tom is not involved in any rules, but the 11am spot is. The 11am time slot is one of the two spots where Yasser can have an appointment. Thus, if Tom's appointment is at 11am, then we know that Yasser's appointment must be at 4pm.

	9	10	11	1	2	3	4	
			Y				Y	UX or XU
	~~S~~ ~~X~~	~~S~~					~~W~~	W...V
2)			(T)				(Y)	

Answer choice (A) is incorrect because if Ulf's appointment is at 1pm and Sam's is at 2pm, then it would be impossible for Xavi and Ulf to have consecutive appointments.

Answer choice (C) is incorrect because Walt cannot have the 3pm appointment, because Victor would have to take the 4pm time slot, which is already occupied by Yasser.

Answer choices (D) and (E) are incorrect because Yasser must have the 4pm appointment.

3. **(A) Walt's appointment is at 11am and Ulf's appointment is at 1pm.**

	9	10	11	1	2	3	4	
			Y				Y	UX or XU
	~~S~~ ~~X~~	~~S~~					~~W~~	W...V
3)			(W)	(U)	↖	↗	(Y)	
					S V X			

In answer choice (A), if Walt has the 11am appointment, then Yasser must have 4pm appointment. And if Ulf has the 1pm appointment, then three entities – Sam, Victor, and Xavi – would have to go into the 2pm and 3pm slots. Thus, answer choice (A) cannot be true.

4. **(B) Walt has an appointment sometime before Sam.**

This question has Ulf in the appointment that is exactly one hour before Victor's appointment. Ulf is involved in the rule that his appointment is either right before or right after Xavi's. Victor is involved in the rule that his appointment is sometime after Walt's. If Ulf has an appointment exactly one hour before Victor, then Xavi must have the time slot one before Ulf. Accordingly, Xavi, Ulf, and Victor form a block of 3 – XUV. Further, Walt has to come sometime before this block of 3, leaving the following possible scenarios:

	9	10	11	1	2	3	4	
			Y				Y	UX or XU
	~~S~~ ~~X~~	~~S~~					~~W~~	W...V
4)	W	X	U	V	S	T	Y	
	W/T	T/W	X	U	V	S	Y	
	T/W	W/T	S	X	U	V	Y	
	T/W	W/T	Y	S	X	U	V	

From these scenarios, you can see that the only answer choice that MUST be true is (B), that Walt has an appointment before Sam.

5. **(E) Victor, Tom, Sam**

Walt has the 11am appointment in this question. Walt is involved in the rule that he has an earlier appointment than Ulf. The 11am appointment is involved in the rule with Yasser. Thus, if Walt has the 11am appointment, then Yasser must have the 4pm appointment. This results in Victor and Sam having to go sometime in between Walt and Yasser. Thus, the Ulf/Xavi pairing cannot go after Walt, as there would be four entities that need to be in the three spots – 1pm, 2pm, and 3pm. Accordingly, the U/X pairing must be in the 9am and 10am slots. Victor, Tom, and Sam are left as the only patients that can be scheduled at 2pm.

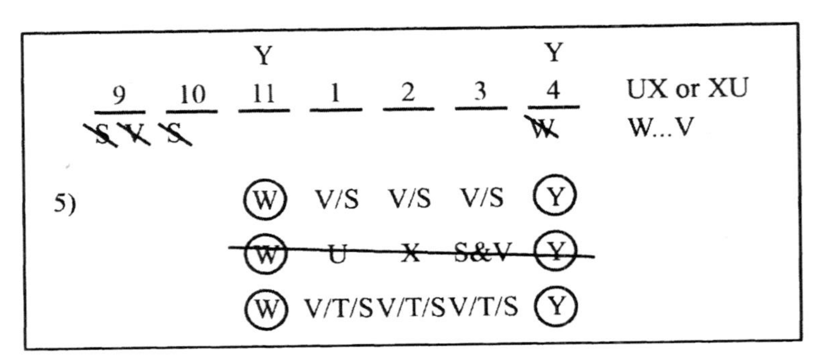

Supermodels

Assigned Ordering GD: Order Entities **Difficulty = Very Easy**

This is a simple assigned ordering game in which you must determine the order of supermodels (entities). The initial setup is simple and straightforward:

```
                                   ABCDEF
_1_  _2_  _3_  _4_  _5_  _6_  _7_
                                    (CD)
                                   A _ _ E
                                    C...I
```

Casablanca is involved in two rules, and so those rules can be combined to deduce that Dominique will also be sometime before the Intermission. Additionally, since this is an assigned ordering game, further deductions can be made as to the first and last spots. Namely, Casablanca cannot be in the sixth or seventh time slots; Dominique cannot be in the first or seventh time slots; and the intermission cannot be in the first or second time slots. Thus, your final diagram will look as follows:

Master Diagram

```
                                   ABCDEF
~D~ ~I~ ~I~                 ~C~ ~C~ ~D~
_1_  _2_  _3_  _4_  _5_  _6_  _7_
                                    (CD)...I
                                   A _ _ E
```

---

1. **(B) Dominique**

This question has Ennui walking fifth. Ennui is involved in the rule that she and Anka must have two spots between them. If Ennui walks fifth, then Anka must walk second (as there are not enough slots after slot 5 for Anka). Since Casablanca and Dominique walk consecutively, but be before the intermission, they must be in time slots 3 and 4 respectively:

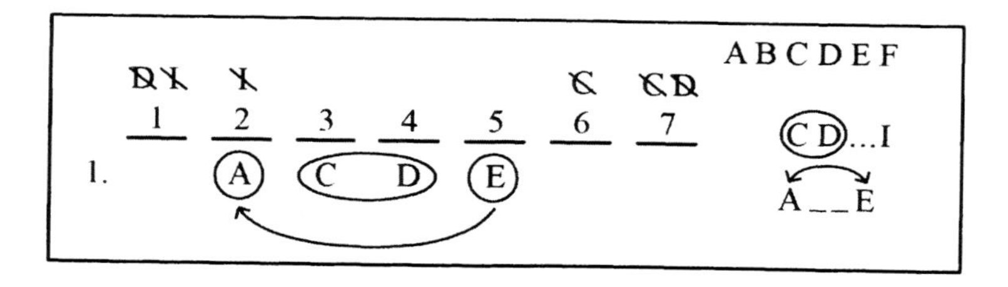

2. **(A) Second**

Recall from the master diagram that Casablanca can never walk sixth or seventh, which allows you eliminate answer choices (D) and (E) immediately. This question has Anka walking first. Anka is involved in the rule that has her exactly three spots away from Ennui. Thus, if Anka walks first, Ennui must walk three spots after in the fourth time slot. This means Casablanca and Dominique can only be consecutive if Casablanca is in the second or fifth time slots.

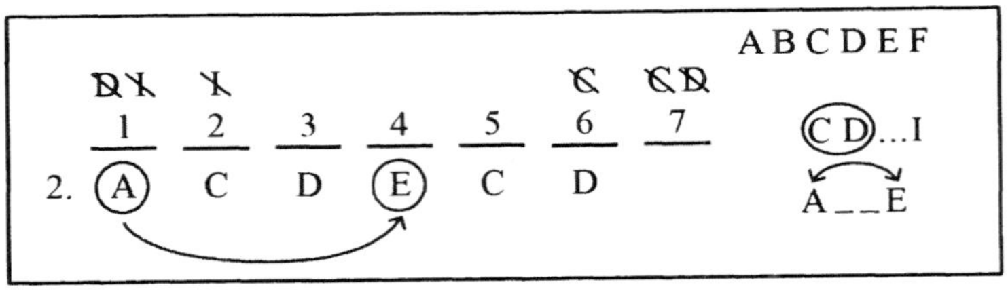

	1	2	3	4	5	6	7
	~~B~~ ~~I~~	~~I~~				~~C~~	~~C~~ ~~B~~
2.	(A)	C	D	(E)	C	D	

ABCDEF

(CD)...I

A _ _ E

3. **(E) Bianca, Anka, Casablanca**

This question has Bianca and Fromage as far apart from each other as possible. Bianca and Fromage are not involved in any rules. Thus, we can start by placing one of the two first and one seventh. Next, we must place the Anka and Ennui in the frame pursuant to the rule that they must have two spots between them. If Bianca and Fromage are first and seventh, then the Anka and Ennui pairing can only be in spots two/five or three/six. However, if Anka and Ennui are in spots three and six, the CD…I grouping will not fit. Thus, we can deduce that Casablanca must go in spot three (and Dominique must go in spot four).

Any answer choices that do not have Casablanca in spot three can be eliminated – answer choices (B), (C), and (E) can be immediately eliminated. Answer choice (A) can also be eliminated as neither Bianca nor Fromage are first.

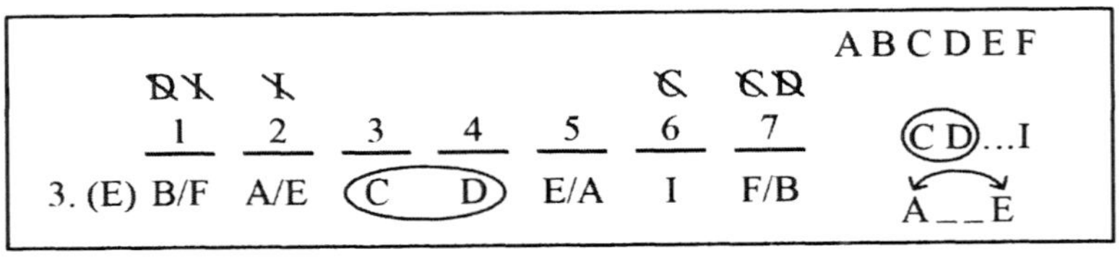

	1	2	3	4	5	6	7
	~~B~~ ~~I~~	~~I~~				~~C~~	~~C~~ ~~B~~
3. (E)	B/F	A/E	(C	D)	E/A	I	F/B

ABCDEF

(CD)...I

A _ _ E

4. **(A) Second**

Per the master diagram, intermission can never be scheduled for the first or second time slots.

5. **(C) Fifth**

Per the master diagram, Casablanca cannot walk sixth or seventh. Casablanca is not directly implicated in any of our other rules or deductions, so the next step is to try out Casablanca in the next latest slot, which is fifth.

If Casablanca walks fifth, then Dominique must walk sixth and intermission is seventh. That leaves the first four slots available. Anka and Ennui must have two slots in between them, so the first slot and the fourth slot will be occupied by either Anka or Ennui. That leaves the second and third slots for Bianca and Fromage. This configuration violates no rules, so the correct answer is (C).

6. **(E) 5**

Since this question is seeking the maximum number of spots between Fromage and the intermission, you should start by placing them as far apart as possible, in slots 1 and 7. The intermission is in slot 7, because it must have at least two supermodels before it – Casablanca and Dominique. Fromage will be placed in slot 1.

Next, we must place the Anka and Ennui in the frame pursuant to the rule that they must have two slots between them. If Fromage is first and the intermission is seventh, then the Anka and Ennui pairing can only be in slots 2/5 or 3/6. Casablanca and Dominique can go in between Anka and Ennui either way without violating any rules. Thus, Fromage and intermission can be in slots 1 and 7 without violating any rules:

ABCDEF

	~~D~~ ~~C~~	~~C~~				~~C~~	~~C~~ ~~D~~
	1	2	3	4	5	6	7
6.	(F)	E/A	(C	D)	A/E	(B)	(I)
	(F)	(B)	E/A	(C	D)	A/E	(I)

(CD)...I

A _ _ E

7. **(D) Fifth**

From Question #2, we learned that Dominique could walk either third or sixth when Anka walks first, so answer choice (A) can be eliminated immediately. The configurations for answer choices (B), (C), (D) and (E) are tested below, to see if Dominique must go in one spot.

ABCDEF

	~~D~~ ~~C~~	~~C~~				~~C~~	~~C~~ ~~D~~
	1	2	3	4	5	6	7
7 B.	C	D	A	C	D	E	I
7 C.	E	C	D	A	C	D	I
7 D.		E	(C	D)	A		I
7 E.	C	D	E	C	D	A	I

(CD)...I

A _ _ E

Per the diagram, answer choice (D) is correct, as Anka in spot five would require Ennui to be in spot two. This would leave only spots three and four for the Casablanca-Dominique pairing.

Suspects in a Lineup

Matching and Selection GD: Match Entities **Difficulty = Easy**

Although the numbers might indicate that this is an assigned ordering game, it operates more like a matching game, as there is not a strong temporal element. The only rule that requires you to "order" is the Ned-Ralph rule that requires those two suspects to be in spots next to each other. Otherwise, you will just be matching entities to spots. In addition, you will select five of the six suspects to place in the lineup.

Master Diagram

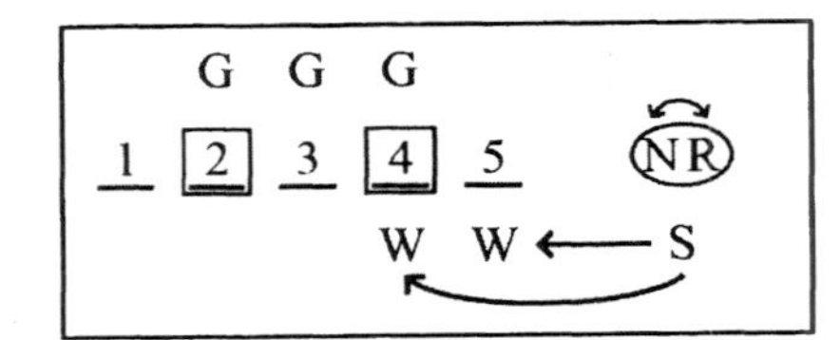

One key deduction in this game are that Ned and Ralph must always been in the lineup because they are always next to each other (and five out of six suspects are picked). The other key deduction is that Willie must always be in the lineup, because if he is not in the lineup then all entities, including Shady, will be in the lineup. But, we know that if Shady is in the lineup then Willie is in the lineup.

1. **(B) Guilty Gabe is in spot 2.**

Shady is involved in the rule that if he is in the lineup then Wylie Willie is in spots 4 or 5. Thus, if Shady is in spot 5, Willie must be in spot 4. Since Ned and Ralph must also be in the lineup, and next to each other, we can then deduce that one of them *must* be in spot 2 while the other can be in spots 1 or 3.

G G G
1 2 3 4 5 NR
W W ← S
1) N/R W S

Since only Ned or Ralph can go in spot 2, Guilty Gabe cannot be in spot 2, and answer choice (B) is correct.

2. **(C) No-Good Ned, Rough Ralph, Shady**

Answer choices (B) and (D) can be eliminated immediately because there are violations in the first three spots themselves. Answer choice (B) has Shady in the lineup, but Wylie is not in spots 4 or 5. Answer choice (D) has Ned in spot 1, but does not have Ralph next to him in spot 2.

Answer choices (A), (C), and (E) must be tested.

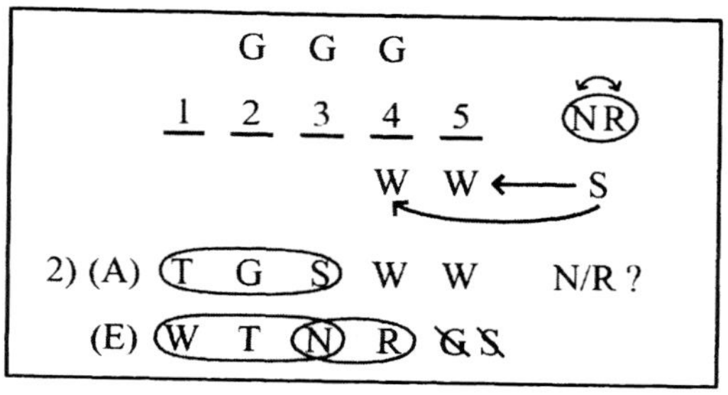

G G G

1 2 3 4 5 (NR)

W W ← S

2) (A) (T G S) W W N/R ?

(E) (W T (N) R) G S

The preceding diagram shows that answer choice (A) is incorrect, because if Shady is in the lineup then Willie would have to be in spots 4 or 5. However, this would not leave room for the Ned/Ralph pairing.

Answer choice (E) shows that in this configuration either Gabe or Shady must be in spot 5. However, this is not possible, as Gabe can only be in spots 2, 3, and 4, and Shady cannot be spot 5 unless Willie is in spot 4.

3. **(D) No-Good Ned, Rough, Ralph, Shady**

There is no quick way to answer this question besides trying out the various configurations with Two-Bit as the criminal.

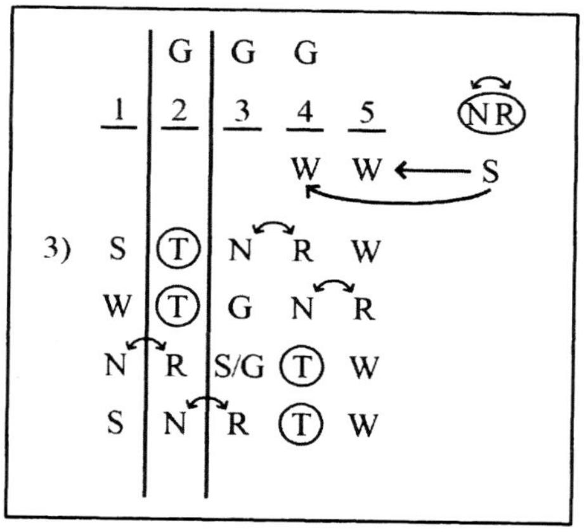

G | G G

1 | 2 | 3 4 5 (NR)

W W ← S

3) S | (T) | N R W

W | (T) | G N R

N | R | S/G (T) W

S | N | R (T) W

The preceding diagram shows that only Two-Bit, Ned, and Ralph can be in spot 2 if Two-Bit is the criminal.

4. **(C) Shady is not in the lineup.**

Answer choices (A) and (B) cannot be true per the deductions that were previously made. The rest of the answer choices must be tested. As always, remember to check prior work to see if the answers have been previously tested.

In this case, answer choices (C) and (E) were both tested in Question #3. In Question #3 we saw configurations that worked that did not have Shady in the lineup. Thus, answer choice (C) is the right answer. To be sure we can also eliminate answer choice (E) with Question #3 because Gabe is not always in the lineups.

Lastly, answer choice (D) can be eliminated as the below diagram shows a configuration that works without Two-Bit in the lineup.

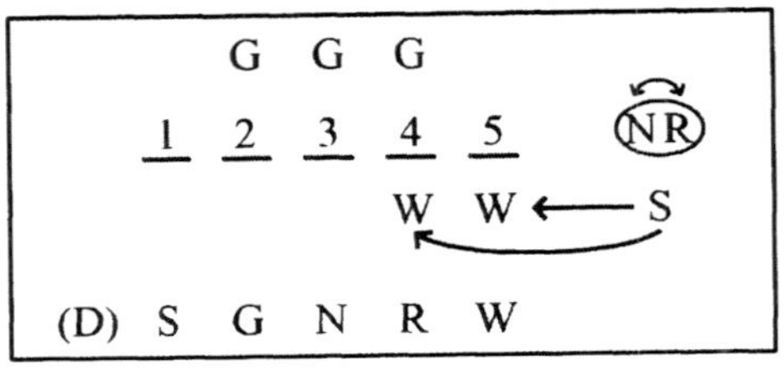

G G G
1 2 3 4 5 (NR)
W W ← S

(D) S G N R W

5. **(E) Wylie Willie, Guilty Gabe, Rough Ralph, No-Good Ned, Two-Bit**

This is essentially just an acceptability question that requires Ned to be in spots 2 or 4.

Answer choice (A) is incorrect because Gabe can only be in spots 2, 3, and 4.

Answer choice (B) is incorrect because Ned is spot 3, which is not a "criminal" spot.

Answer choice (C) is incorrect because Ned and Ralph are not next to each other.

Answer choice (D) is incorrect because Shady is in the lineup, but Willie is not spots 4 or 5.

Geography Lesson

Assigned Ordering GD: Order Entities **Difficult = Easy**

In this assigned ordering games, regions of the world (entities) will be scheduled in different months (spots). There are several positional rules, and only one relational rule, which helps to make the diagram, and the game as a whole, fairly easy.

Master Diagram

Af, As, Eur, SA, NA, US

As	As					
Sep	Oct	Nov	Dec	Jan	Feb	Eur __ US
~~US~~	~~US~~	~~NA~~ Af	⟶	SA or	SA	
				~~Eur~~	~~Eur~~	

1. **(C) Africa, Asia, Europe, South America, United States, North America**

Answer choice (A) is incorrect because North America cannot be taught in November.

Answer choice (B) is incorrect because Europe has to be two spots before United States.

Answer choice (D) is incorrect because when Africa is taught in November, South America has to be taught during the last two months.

Answer choice (E) is incorrect because Asia has to be taught in September or October.

2. **(E) United States is taught in a later month than Africa.**

This question has South America taught in September. South America is involved in the rule that states it must be taught in January or February if Africa is taught in November. Thus, Africa cannot be taught in November, as the question requires that South America is taught in September. The September spot is involved in the rule that requires Asia to be taught in September or October. Thus, Asia is taught in October.

For the month of November, only Europe can be taught. United States cannot be taught in November, because that would mean Europe would have to be taught in September, and September is already occupied by South America. The following diagram results:

	As	As					
	Sep	Oct	Nov	Dec	Jan	Feb	Eur __ US
	~~US~~	~~US~~	~~NA~~ Af	⟶	SA or	SA	
					~~Eur~~	~~Eur~~	
2)	SA	As	Eur	Af/NA	US	Af/NA	

Based on the above, answer choice (E) is the only one that could be false.

3. **(B) 2**

This question has Africa taught in November and asks which months Europe can be taught. From the master diagram, we know Europe can never be taught on January or February. If Africa is taught in November, then Europe also cannot be taught in November. Further, Europe cannot be taught in September, because this would leave the United States in November. Thus, Europe can only be taught in October or December in this scenario.

	As	As					
	Sep	Oct	Nov	Dec	Jan	Feb	Eur __ US
	~~US~~	~~US~~	~~NA~~ Af	⟶	SA or	SA	
					~~Eur~~	~~Eur~~	
3)	~~Eur~~		(Af) ~~Eur~~		SA~~Eur~~	SA~~Eur~~	

4. **(D) South America is taught before United States.**

This question has United States taught in November. The United States is involved in the rule that it must be two spots after Europe. November is involved in the rule that North America cannot be taught in November. If the United States is taught in November, Europe would have to be taught in September and Asia in October, resulting in the following diagram:

	As	As					
	Sep	Oct	Nov	Dec	Jan	Feb	Eur __ US
	~~US~~	~~US~~	~~NA~~ Af	⟶	SA or	SA	
					~~Eur~~	~~Eur~~	
				Af	Af	Af	
4)	(Eur)	(As)	(US)	SA	SA	SA	
				NA	NA	NA	

Based on the above, only answer choice (D) must be false.

5. **(A) January, February**

This question has North America taught in between Europe and United States creating a Eur-NA-US block. North America is involved in the rule that it cannot be taught in November. Europe and United States are involved in the rule that Europe must be taught two months before the U.S.

Per the master diagram, United States can never be taught in September or October. Further, since North America must be taught in between Europe and United States, we know that United States cannot be taught in December either, as this would result in North America in November, which is not allowed. Lastly, if United States is taught in November, then North America would be taught in October, and Europe in September. This would leave no months for Asia to be taught.

Thus, the only months United States can be taught if North America is taught directly before, are January and February. Answer choice (A) is the correct answer.

Wine and Cheese Pairing

Selection and Matching GD: Match Entities **Difficulty = Moderate**

This game has a lengthy introduction with rules tucked into the background. However, a thorough and meaningful reading of the background and rules reveals a manageable selection game that requires you to select wines and cheeses and match them. Since this is primarily a selection game, there is no frame, and the diagram is straightforward:

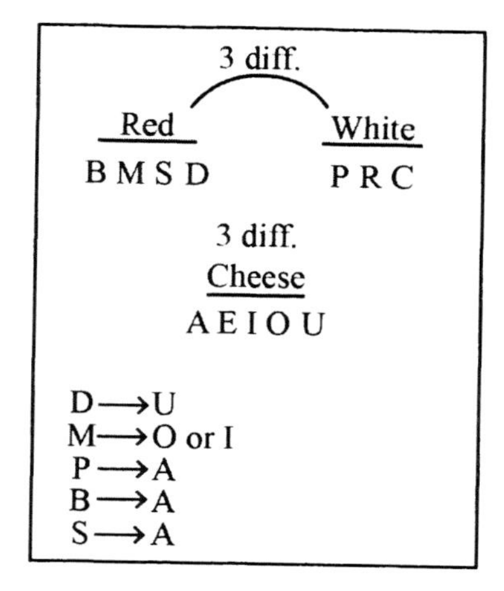

The rule that Pinot Gris, Bordeaux, and Shiraz all pair with Ambert can be combined with the rule that each flight has three different cheeses (remember not to forget the rules introduced in the opening paragraph), resulting in the deduction that that Pinot Gris, Bordeaux, and Shiraz can never be offered together:

P⟶A
B⟶A ⟶ ~~PB~~ ~~BS~~ ~~PS~~
S⟶A

Putting it all together results in the following master diagram:

Master Diagram

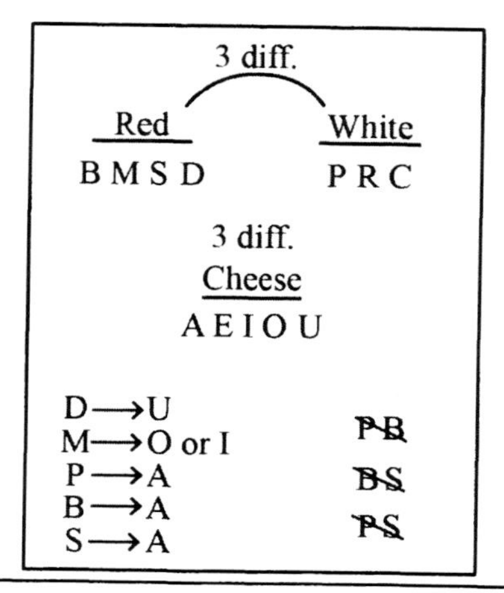

1. **(D) Ambert, Idiazabal, Ubriaco**

An All-Red Flight would have any three of the four the red wines – Bordeaux, Malbec, Shiraz, or Dolcetto. However, Bordeaux and Shiraz cannot be offered together per our deductions. Thus, Malbec and Dolcetto must be served together during Red Flights (deduction!). The other wine can be either Bordeaux or Shiraz.

Since Dolcetto must always be paired with Ubriaco, we know that any answer that does not include Ubriaco must be wrong, and you can eliminate answer choices (A), (B), and (C). Further, we know that Bordeaux or Shiraz have to be paired with Ambert. Thus, answer choice (E) is incorrect, which leaves (D) as the correct answer.

2. **(D) The flight is White or Mixed.**

This question has Pinot Gris and Emmental served. Pinot Gris is involved in a rule that requires it to be paired with Ambert. Emmental is not involved in any rules, but it cannot be paired with D, M, P, B, and S, since those wines already have specified pairings. Thus only two white wines, Riesling and Chardonnay, are left for Emmental to pair with. Accordingly, we know that there are at least two white wines selected, and that answer choice (D) is correct.

3. **(D) Emmental and Idiazabal are served.**

The European Wines are Bordeaux, Dolcetto, Pinot, and Riesling – two reds and two whites. Bordeaux is involved in a rule that requires it to be paired with Ambert. The same goes for Pinot, it must be with Ambert. Thus, only one of those two can be selected. This means Dolcetto and Riesling MUST be selected in order for there to be three wines on the flight. Dolcetto is involved in a rule that requires it to be paired with Ubriaco.

Accordingly, Ubriaco and Ambert must be served. Answer choice (D) is the only that does not allow for this, as neither Ambert nor Ubriaco appear in the answer.

4. **(C) Malbec, Bordeaux, Dolcetto**

This question has Oaxaca and Idiazabal served and can be answered easily by elimination. Oaxaca and Idiazabal are involved in the rule with Malbec, in which Malbec must be paired with either of the two. If Malbec is offered, it will be paired with one of the two, and one of the other two wines would have to be paired with whichever of Oaxaca or Idiazabal is left. Thus, at least one of the wines offered must not be involved in a rule (so that it is free to take on Oaxaca or Idiazabal). Riesling and Chardonnay can pair with any of the cheeses.

Only answer choice (C) does not allow for this, as Bordeaux and Dolcetto already have cheeses they must be paired with. Therefore, (C) is the correct answer.

5. **(C) The flight is Mixed.**

Ambert is not served in this question. Ambert is involved in the rule that pairs it with Pinot, Bordeaux, and Shiraz. If Ambert is not served, then neither are Bordeaux and Shiraz – reds – nor is Pinot – a white. Thus, this ends up being a very simple question, as there cannot be a full Red or full White flight in this situation.

6. **(E) Wines: Pinot Gris, Riesling, Chardonnay**

Cheeses: Ambert, Emmental, Idiazabal

Answer choice (A) is incorrect because Shiraz must be served with Ambert.

Answer choice (B) is incorrect because Ubriaco must be served with Dolcetto.

Answer choice (C) is incorrect because Ambert cannot be served twice.

Answer choice (D) is incorrect because Pinot Gris and Shiraz cannot be served together, as both are paired with Ambert.

Russian Composers

Relative Ordering	GD: Order Entities	**Difficulty = Moderate**

The only potential nuisance in this relative ordering game is that some of the composers (entities) have two pieces. The two pieces can be designated by subscripts. You should also notice that Obukhov is the only entity with no restrictions and therefore is free floating in this game.

O P_1 P_2 R_1 R_2 S T_1 T_2 U

```
P
 ·.: R1...P2
U.:'
   '·S...T1...T2
```

As with all relative ordering games, once the diagram is drawn you should then determine which of the entities could be first and which could be last. In this case, only the first Prokofiev, Ustvolskaya, and Obukhov have no pieces that must come before them. Further, only the second Tchaikovsky, the second Prokofiev, the second Rachmaninoff, and Obukhov have no pieces that must come after them.

Master Diagram

```
P                          1    9
 ·.: R1...P2               O    O
U.:'                       P    P
   '·S...T1...T2           U    R
                                T
```

1. **(D) Prokofiev, Ustvolskaya, Rachmaninoff, Stravinsky, Tchaikovsky, Tchaikovsky, Prokofiev, Rachmaninoff, Obukhov**

Answer choice (A) is incorrect because Ustvolskaya must be played before Stravinsky.

Answer choice (B) is incorrect because Stravinsky must be played before Tchaikovsky.

Answer choice (C) is incorrect because only Obukhov, Prokofiev, and Ustvolskaya can be first.

Answer choice (E) is incorrect because the first Rachmaninoff must come before the second Prokofiev.

2. **(D) Stravinsky**

This is a very simple question if you have diagrammed correctly. As per the master diagram, only the second Tchaikovsky, the second Prokofiev, the second Rachmaninoff, and Obukhov can be last.

3. **(E) Tchaikovsky**

If Ustvolskaya is second, the master diagram shows us that the second Prokofiev and both Tchaikovskys cannot be third because the first Rachmaninoff goes in between Ustvolskaya and the second Prokofiev, and the Stravinsky goes between Ustvolskaya and both Tchaikovskys.

1	2	3
P/O	U	~~P_2~~ ~~T_1~~ ~~T_2~~

Thus, answer choices (A), (C), and (D) can be eliminated. Answer choice (E) is correct here because answer choice (B) does not specify which of the Prokofievs is third. The first Prokofiev can certainly be played third, even though the second Prokofiev cannot, making answer choice (B) incorrect.

4. **(A) Tchaikovsky is played fourth.**

This question has second Rachmaninoff piece played between the two Tchaikovsky pieces. The key to this question is recalling that Obukhov has no restrictions and can be placed in any spot. Putting Obukhov in any of the first five spots allows a Tchaikovsky to be played in spots 5, 6, 7, and 8:

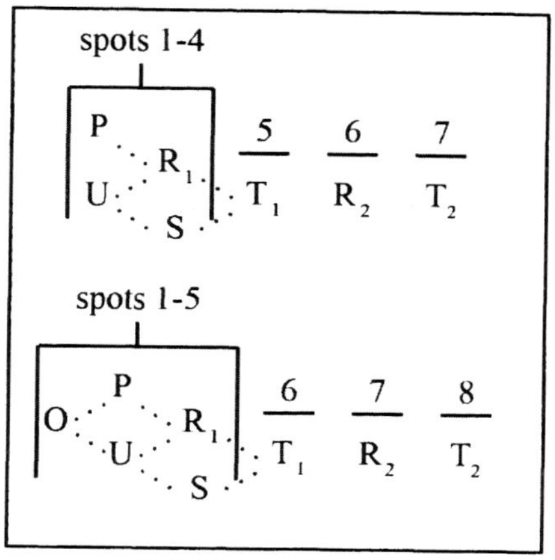

spots 1-4

P, U, R_1, S

5	6	7
T_1	R_2	T_2

spots 1-5

P, O, U, R_1, S

6	7	8
T_1	R_2	T_2

The preceding diagram shows that the earliest the first Tchaikovsky can be played is fifth, and so answer choice (A) is correct.

5. **(C) Rachmaninoff is played seventh and eighth.**

This question has the Prokofiev pieces first and last. The question is not easy to solve by just looking at the master diagram, so each answer choice should be diagrammed to see if that answer choice violates any rules.

	1	2	3	4	5	6	7	8	9
	Ⓟ								Ⓟ
(A)	Ⓟ		Ⓞ	Ⓤ	S	T_1	T_2	R_1	Ⓟ
(B)	Ⓟ		Ⓤ	Ⓞ	S	T_1	T_2	R_1	Ⓟ
(C)	Ⓟ	O	U	S	T_1	T_2	(R_1)	(R_2)	Ⓟ
(D)	Ⓟ						Ⓢ	Ⓣ	Ⓟ
(E)	Ⓟ						Ⓢ	Ⓞ	Ⓟ

Answer choice (A) and (B) are incorrect because there is no room for the second Rachmaninoff.

Answer choices (D) and (E) are incorrect because there is no room for the Tchaikovskys.

6. **(C) Obukhov, Prokofiev, Stravinsky, Tchaikovsky**

Answer choice (C) is correct because Ustvolskaya must always come before Stravinsky.

Desert Safari

Positioning GD: Position Entities **Difficulty = Moderate**

This game requires you to position seven friends (entities) in various seats (spots) in an SUV. The game itself is quite descriptive, lending to a diagram that is very easy to set up:

Master Diagram

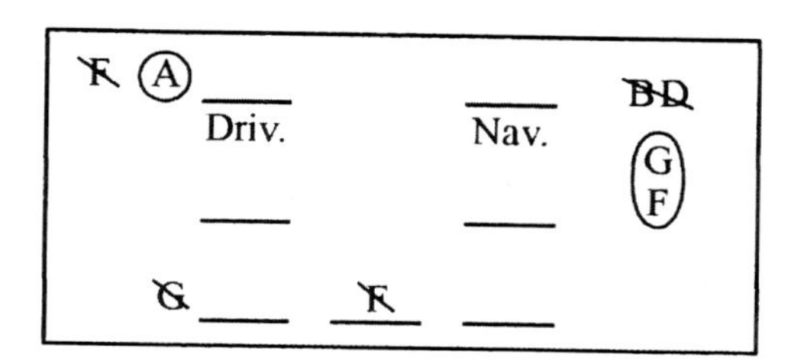

Farhan cannot be in the first row or in the middle seat of the back row, as Grace must always be directly in front of him. These deductions were drawn directly into the diagram.

1. **(E) Esam is in the back row.**

This question has Chamis as the Navigator and Grace behind him. If Chamis is the Navigator, Aisha must be the Driver, as she can only be the Driver or Navigator. If the Grace sits behind Chamis, then Farhan must be in the back right seat. Further, since neither Bilal nor Diyaa can sit next to each other, we can deduce that one of them must sit in the middle row behind the Driver.

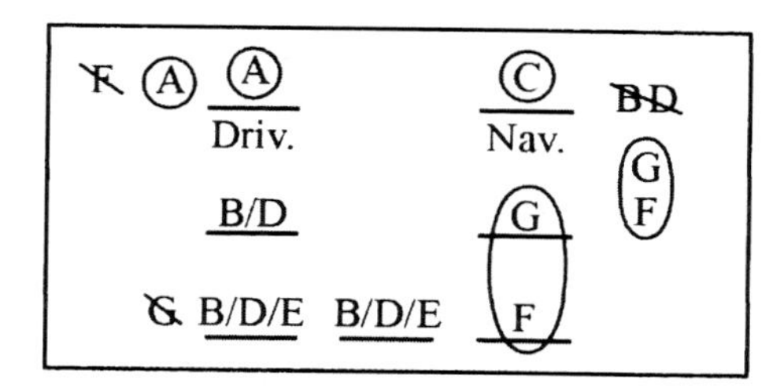

Per the preceding diagram, Esam must be in the back row.

2. **(B) Farhan**

This question is easily answered by our master diagram. Farhan cannot be in the back middle seat because Grace has to be directly in front of him, and there are no seats in front of the back middle seat.

3. **(C) Grace is the Navigator.**

If Aisha is the Navigator and Esam sits behind her then the Grace/Chamis pairing must be on the right side of the car. However, if Grace is the Navigator and Farhan is the middle row, that would leave Bilal and Diyaa in the same row, which is not allowed.

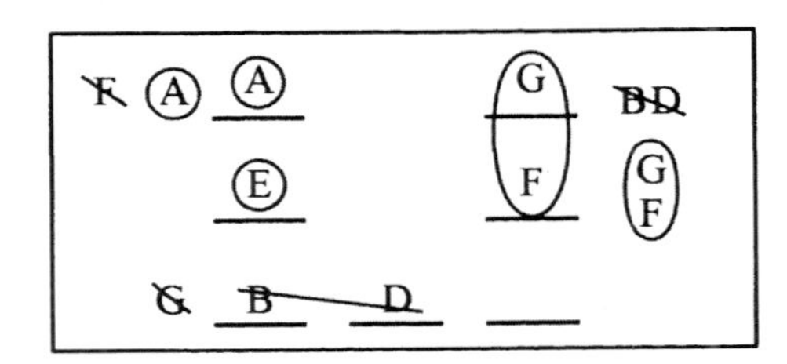

Thus, Grace must be in the middle row behind the Navigator. This will put Farhan in the back row and either Bilal or Diyaa will be the Navigator. Further, Chamis must be in the back row.

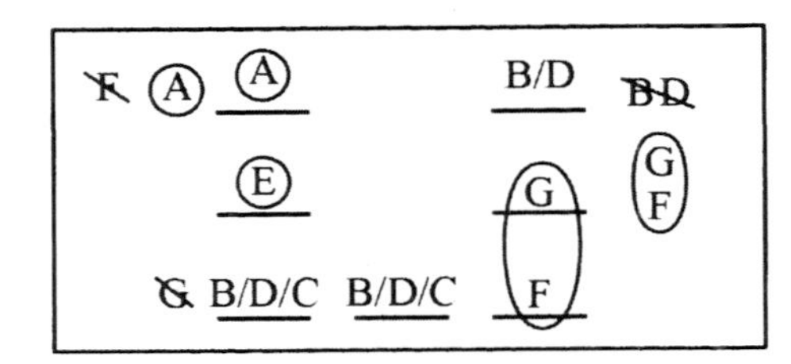

Thus, each of the answer choices except (C) could be true.

4. **(D) Chamis is the driver.**

If Grace is not next to Bilal or Diyaa then two possible configurations result:

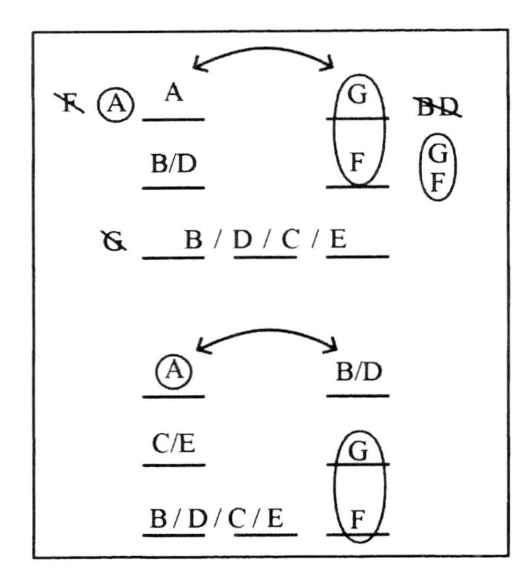

In both of the preceding configurations the right and left rows are interchangeable. Thus, answer choices (B) and (C) can be eliminated although the diagram does not exactly show this. See Chapter 9 for more discussion on the special discussion that allowed us to come to the conclusion.

To be sure, Grace is either in the middle row or the first row, meaning that answer choice (A) could be true or false, so answer choice (A) can be eliminated. Based on the diagram, Chamis could be in the middle or back row, but not in the front row. Answer choice (E) can be eliminated and answer choice (D) is the correct answer.

5. **(D) Chamis, Esam, Farhan**

This is a straightforward acceptability question, which can be answered solely by using the master diagram.

Answer choice (A) is incorrect because Bilal and Diyaa cannot be in the same row.

Answer choice (B) is incorrect because Grace cannot be in the back row.

Answer choice (C) is incorrect because Farhan cannot be in the middle seat of the back row.

Answer choice (E) is incorrect because Aisha can only be in the front.

Pirate Plank Walk

Assigned Ordering GD: Order and Position Entities **Difficulty = Moderate**

This game requires you to order and position pirates (entities) in a line. The pirates will then be sent, one at a time, to walk the plank. The twist to this game is that there are two sets of spots, as the pirates can be sent in order from left to right or from right to left.

The rules dictate the position in which the pirates stand *only* and *not the order they walk the plank.* The questions will then determine whether the pirates are sent to walk left to right or right to left. Arrows in the sidebar can be used to depict whether the pirates are walking right to left or left to right.

Master Diagram

6	[5]	4	3	2	1	
1	2	3	4	[5]	6	(BG)
(C)	~~D~~				~~B~~	

Remember the deduction that Bluebeard can never be furthest to the right as GreenBeard must always be directly to the right of him. Combining this with the fact that Calico Jack is always furthest to the left results in the deduction that BlueBeard can never be first or last to walk the plank!

1. **(C) Diabolito walks the plank second.**

This question has Long John walking the plank fourth. The following diagram shows the configuration where Long John walks fourth going right to left and the configuration where Long John walks fourth going left to right.

	6	[5]	4	3	2	1	
	1	2	3	4	[5]	6	(BG)
	(C)	~~D~~				~~B~~	
1)	(C)		(L)		B/G		←
	(C)			(L)			→

The only known spots going right to left are spots 6, which has to be Calico Jack, spot 4, which has to be Long John, and spot 2, which has to be either BlueBeard or GreenBeard. The only known spots going left to right are spot 1, which has to be Calico Jack, and spot 4, which has to be Long John.

Based on the above, only answer choice (C) must be false, as Diabolito can never be in spot 2 going left to right, per the master diagram, nor can he be in spot 2 to going right to left, because either BlueBeard or GreenBeard must be in spot 2.

2. **(E) GreenBeard is granted Parley and the order is from left to right.**

If Eric is immediately to the left of Long John, then two pairings exist. Diabolito cannot be next to Calico Jack and cannot split the two duos of pirates, yielding four configurations:

	6	5	4	3	2	1	
	1	2	3	4	5	6	
	C	~~D~~				~~D~~	BG
2)	C	E	L	D	B	G	
	C	E	L	B	G	D	
	C	B	G	D	E	L	
	C	B	G	E	L	D	

Answer choice (A) is incorrect because when the order is left to right, only Long John and GreenBeard can be third.

Answer choice (B) is incorrect because when the order is right to left, only Diabolito, BlueBeard, and Eric can be third.

Answer choice (C) is incorrect because Diabolito cannot be fifth in either configuration.

Answer choice (D) is incorrect because when the order is right to left, only Eric and BlueBeard can be fifth.

3. **(B) GreenBeard, Eric the Red, Long John**

Since the question states BlueBeard is next to the pirate who is in spot 5, he must be in either spots 4 or 6. BlueBeard is also involved in the rule that requires GreenBeard to always be to his immediate right.

If the order is right to left, BlueBeard cannot be in spot 6, because Calico Jack is in that spot already. Thus, he must be spot 4. This leaves only Eric the Red and Long John for spot 5.

If the order is left to right, BlueBeard cannot be in be in spot 6, per the master diagram, and so he must be in spot 4. If BlueBeard is in spot 4 then GreenBeard *must* be in spot 5. This means either Eric the Red, Long John, or GreenBeard could be granted Parley.

	6	5	4	3	2	1	
	1	2	3	4	5	6	
	C	~~D~~				~~D~~	BG
3)	C	E/L	B	G			←
	C			B	G		→

4. **(D) GreenBeard walks the plank second.**

Answer choice (D) is not true when going left to right, as this would require BlueBeard to be in spot 1, which is already occupied by Calico Jack.

	6	5	4	3	2	1	
	1	2	3	4	5	6	
	(C)	~~D~~				~~B~~	(BG)
4) (A)	(C)	(L)	(B	G)	E	D	
	(C)	E	(B	G)	(L)	D	

Answer choice (B) can be true, as the diagram from Question #2 shows.

Answer choice (C) can be true, as the diagram from Question #2 shows.

Answer choice (E) can be true, as the diagrams from Questions #2 and #3 show.

5. **(A) BlueBeard**

We made the key deduction that BlueBeard cannot be first to walk the plank when setting up the master diagram. Going left to right he cannot be in spot 1 because Calico Jack is furthest to the left. Going right to left he cannot be in spot 1 because GreenBeard is immediately on his right.

6. **(D) First: Diabolito; Second: Eric the Red; Third: GreenBeard**

This question does not state whether the pirates are being sent left to right or right to left, and so you will have to deduce this from the answer choices. Since the question is dealing with the first three spots, if Calico Jack is first we know that the order is left to right. Otherwise, it is right to left.

Answer choice (A) is left to right. It is incorrect because BlueBeard must be second if GreenBeard is third.

Answer choice (B) is right to left. It is incorrect because if GreenBeard is third, BlueBeard must be fourth, and if BlueBeard is fourth then Diablo would end up in spot 5, which is not allowed going right to left.

	6	[5]	4	3	2	1	
	(C)	~~D~~				~~B~~	(BG)
5) (B)	(C)	~~D~~	(B	G)	L	E	←

Answer choice (C) is right to left. It is incorrect because BlueBeard is on the right of GreenBeard instead of on the left of GreenBeard.

Answer choice (E) is right to left. It is incorrect because GreenBeard is not immediately to the right of BlueBeard (Diabolito is to the right of BlueBeard going right to left).

Senate Committees

Matching (Distribution) GD: Match Entities **Difficulty = Hard**

This is a matching game that requires you to place senators (entities) into three committees (spots). There are ten entities and four sets of entities in this game – northern, southern, eastern, and western senators. There are several positional rules that in this game that can be drawn directly into the frame, making this somewhat difficult game a bit easier to handle:

Master Diagram

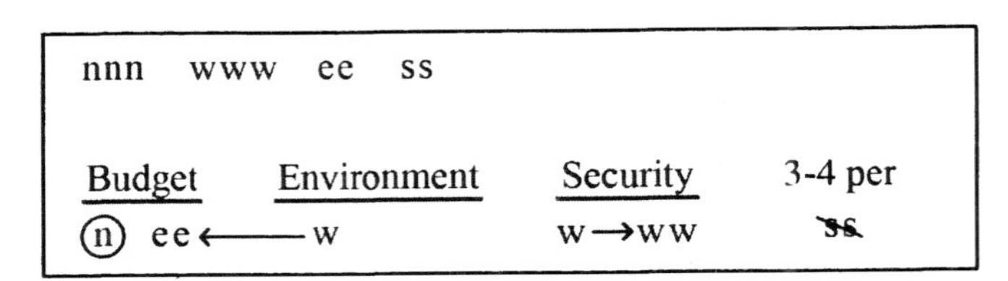

The western senators are involved in two rules, and the Budget committee is involved in two rules. However, the combination of these rules will not result in any significant deductions. The rule that each committee has as least three senators combined with the fact that there are only ten senators yields the deduction that each committee can have, at most, four senators. No further significant deductions can be made so we can move on to the questions.

1. **(E) Exactly three senators serve in the Security committee.**

The Budget committee is involved in the rule that if any western senator is in the Environment committee both eastern senators are in the Budget committee. Thus, if all three northern senators are in the Budget committee we know that no western senator can be in the Environment committee, as that would put both eastern senators in the Budget committee, causing more than four senators to be in the Budget committee (which is not allowed).

Further, no western senator can go in the Budget committee because if one western senator goes in the Budget committee then the others would have to go in the Environment committee or the Security committee. However, no western senators can go in the Environment committee and if one western senator is the Security committee they all must be there.

Budget Environment Security 3-4 per
(n) ee ← w w → ww
1) (nnn) w w

Accordingly, we can deduce that if all the northern senators are in the Budget committee all of the western senators must be in the Security committee. Further, we can also deduce that exactly one of the southern senators must be in the Environment committee because the two southern senators cannot be together, and the Environment committee needs at least three senators.

Further, both of the eastern senators must be in the Environment committee in order for the Environment committee to have three senators. This leaves one southern senator able to serve in either the Budget or the Security committees.

	Budget	Environment	Security	3-4 per
	(n) ee ←	w	w → ww	~~ss~~
1)	(nnn) s	(s)(e)(e)	(www) s	

Answer choices (A), (B), (C), and (D) all must be true. Answer choice (E) could be true, but is not necessarily true.

2. **(E) 10**

If a western senator is in the Environment committee then, per the master diagram, both eastern senators are in the Budget committee. Further, this means that the third western senator cannot be in the Security committee, because if one western senator is in the Security committee, all senators are in the Security committee.

	Budget	Environment	Security	3-4 per
	(n) ee ←	w	w → ww	~~ss~~
2)	(n) (ee)	(ww)	~~w~~	

Per the above, we can deduce that the other western senator must be in the Budget committee, giving the Budget committee four senators, the max allowed. This means one of the southern senators must go in the Environment committee and the other must go in the Security committee. Lastly, the final two northern senators must serve in the Security committee in order for the Security committee to have three senators.

	Budget	Environment	Security	3-4 per
	(n) ee ←	w	w → ww	~~ss~~
2)	(n)(w)(ee)	(ww)(s)	(s)(n)(n)	

Therefore, the committees for all ten senators can be determined.

3. **(D) An eastern senator serves in the Security committee.**

If no southern senator serves in the Environment committee, one must serve in the Budget committee and the other must serve in the Security committee. Further, no western senator can serve in the Budget committee because this would leave the other western senators in either the Environment committee (which would result in both eastern senators in the Budget committee and a total of five senators in Budget) or the Security committee (which is not allowed because if one western senator is in the Security committee they all must be in the Security committee).

The result is two possible combinations – one with all of the western senators in the Environment committee and the other with all the western senators in the Security committee.

	Budget	Environment	Security	3-4 per
	n ee ←	w	w → ww	ss
1)	n s w	s		
2)	n s n/e	n e n/e	s www	
3)	n s ee	www	s n n	

Answer choice (D) is the only one that cannot be true. An eastern senator cannot be in the Security committee in the first configuration, as there are already four senators in the committee. An eastern senator cannot be in the Security committee in the second combination, because if there is a western senator in the Environment committee, then both eastern senators must be in the Budget committee.

4. **(A) A western senator serves in the Environment committee.**

Whenever possible, you should use prior work to save time in answering questions. In this instance, four senators in the Security committee was diagrammed in Question #3 and Question #1.

	Budget	Environment	Security
1)	nnn	see	wwws
3)	ns n/e	ne n/e	s www

Based on these previous diagrams, we can see that answer choices (B), (C), (D), and (E) can all be true. Thus, answer choice (A) is correct.

5. **(D) Budget: north, east, east; Environment: west, west, west, south; Security: north, south, north**

Answer choice (A) is incorrect because if the Environmental committee has a senator from the west, both eastern senators serve in the Budget committee.

Answer choice (B) is incorrect because if a western senator serves in the Security committee, then all senators from the west serve in the Security committee.

Answer choice (C) is incorrect because the two southern senators cannot be in the same committee.

Answer choice (E) is incorrect because the Budget committee must have at least one northern senator.

Symposium Speakers

Relative Ordering GD: Order Entities **Difficulty = Hard**

In this relative ordering game, you must order symposium speakers (entities) comprised of two groups – professors and reverends. As with all relative ordering games you will want to try and combine all of the rules into a tree diagram. Once you finish drawing your tree, make note of which entities can be first and which can be last.

Ickles (P), Lemke (P), Nielsen (R), and Plum (P) are the only speakers that can speak first. Keats (P), Lemke (P), Mitchell (R), and Quintess (R) are the only speakers that can speak last. Lemke is the only "floater" with no restrictions. Putting all of this together yields the following master diagram.

Master Diagram

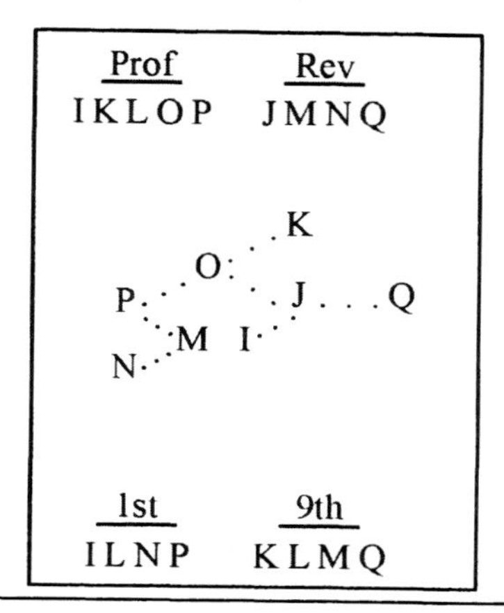

1. **(E) Plum, Lemke, Nielsen, O'Hara, Mitchell, Keats, Ickles, Johnson, Quintess**

Answer choice (A) is incorrect because Ickles must speak before Johnson.

Answer choice (B) is incorrect because O'Hara must speak before Johnson.

Answer choice (C) is incorrect because O'Hara must speak before Keats.

Answer choice (D) is incorrect because Johnson can never speak last.

2. **(C) Three reverends and one professor speak before Keats.**

This question has Keats speaking fifth. Professor Plum and Professor O'Hara must always speak before Keats. Thus, at least two professors must speak before Keats. Inversely, this means that no more than two reverends can speak before Keats. Accordingly, answer choice (C) is correct.

3. **(B) Keats, Lemke, Mitchell, Quintess**

This question is easy to answer from the master diagram. Keats, Mitchell, and Quintess are the only entities that do not have "branches" coming off to the right. This signifies that Keats, Mitchell, and Quintess can potentially be last. Further, Lemke is a floater, with no restrictions, and thus can also be last.

4. **(A) Johnson can speak no earlier than fifth.**

Answer choice (A) is correct because Plum, O'Hara, and Ickles always come before Johnson. Additionally, Lemke comes before Johnson in this question. Thus, there are four speakers before Johnson, making the earliest he can speak as fifth.

Answer choice (B) is incorrect because Ickles can speak as early as second.

Answer choice (C) is incorrect because there are only four speakers that must come before Quintess – Lemke, Plum, O'Hara, and Johnson.

Answer choice (D) is incorrect because Professor Keats can be last.

Answer choice (E) is incorrect because Reverend Quintess and Reverend Mitchell can be last.

5. **(D) 5**

This question has Mitchell speaking fourth, and is the second reverend to speak. If Mitchell is the second reverend, then Nielsen must be the first reverend speaking, and will be in one of the first three spots. Moreover, Mitchell must speak after Plum, per our master diagram. This leaves exactly 1 additional spot available before Mitchell speaks. Accordingly, any speakers that must have more than one other speaker before them cannot go in spot 3.

Johnson, Keats, and Quintess, cannot speak third because they all have at least two speakers that must come before them. Further, Mitchell cannot come third because the question has Mitchell speaking fourth. There are no rules or deductions that restrict the other remaining five speakers from being in spot 3. Thus, four entities/speakers cannot be in spot 3, and the other five entities/speakers can.

Parasailing

Matching & Positioning GD: Match and Position Entities **Difficulty= Hard**

This is a complicated matching and positioning game, which involves the placement of the nine friends (entities) into positions on a boat (spots). The first step diagramming this game is to draw the correct frame. Since this game involves a boat, you should try to imagine a boat with rows when drawing the frame:

```
                guys     girls
                ABCDE    FGHI

(Front)
    __          Pilot
__ __ __        Front Bench
  __ __         Back Bench
    __          Line Keeper
  __ __         Parasail
```

Next, draw in the rules as provided:

```
                guys     girls
                ABCDE    FGHI

    A           E
__ __ __        .
  __ __         F
    A           (FH)
  C/I __        CI (crossed out)
```

Can any of the rules be combined for deductions? There are two rules that involve Frida, so those rules certainly can be combined for the deduction that Eldrick must also sit somewhere in front of Haven. Further, we know that since Frida and Haven must sit next to each other, neither one can be the pilot, the line keeper, nor in the parasail because there is only one spot available in each of those rows.

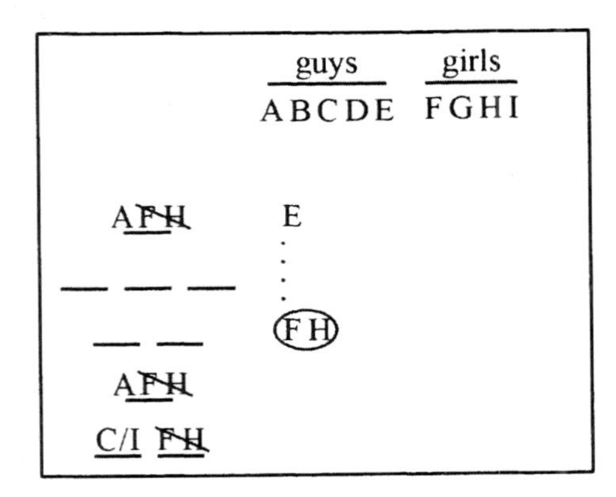

Since Amani has to be either the pilot or the line keeper, and no one sits in the same spot for both trips, in one trip Amani must be the pilot and in the other he must be a line keeper. This results in two possible combinations. The first combination has Amani as a pilot, with Eldrick in the front row and Frida/Haven in the back row (because Eldrick must be in a row in front of Frida/Haven).

The second combination has Amani as the line keeper. Since, Frida/Haven were in the back row in the first combination, they must be on the front row in the second combination. Hence, Eldrick must then be the pilot.

Master Diagram

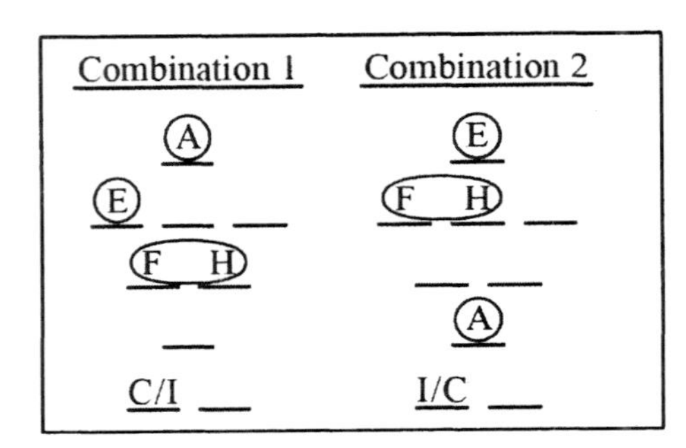

Making the above deductions will significantly reduce the possibilities and help to simplify this otherwise complicated logic game.

---

1. **(B) Amani, Eldrick**

This question tests whether you have made the very important deduction that only Amani and Eldrick can pilot the boat. This type of question is a gift on the LSAT, which you must answer in only seconds in order to put time into the more difficult ones.

2. **(E) Demonde sits in the parasail.**

This question has Bowie and Galilea sitting next to each other on the outbound journey. Bowie and Galilea are not involved in any rules so they are free to sit just about anywhere. If Bowie and Galilea are sitting next to each other, per the question, they can only go in the front row in Combination 1 or in the back row in Combination 2. Taking into account the fact that Irie and Clifton cannot sit on the same row together, the following diagram results.

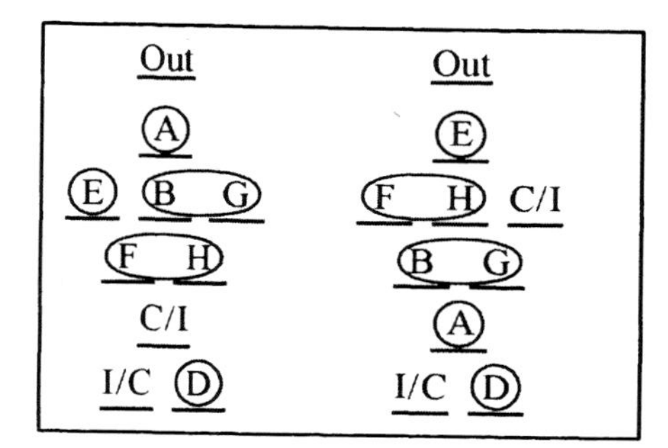

The return diagrams are therefore as follows:

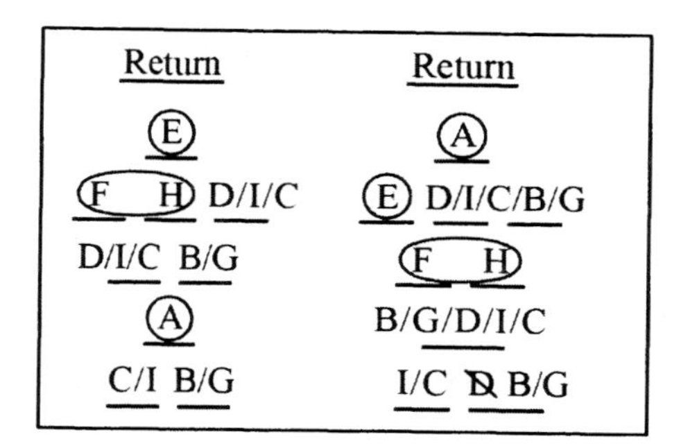

The only answer choice that cannot be true is (E) because Demonde cannot be in the parasail for both journeys.

3. **(C) Clifton is on the back bench on the return journey.**

This question has Bowie in the parasail on the outbound journey and as the line keeper on the return journey. The line keeper is involved in the rule that requires Amani to be either the line keeper or the pilot. Thus, if Bowie is the line keeper on the return journey, Amani is the pilot on the return journey. Hence, the return journey is "Combination 1" and the outbound journey is "Combination 2" from our master diagram. Filling in the known entities from the master diagram, we can see that the back bench must be Frida and Haven.

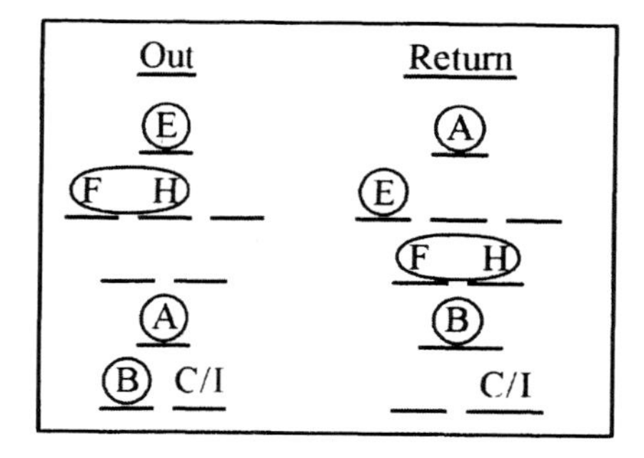

Per the preceding diagram, answer choice (C) cannot be true.

4. **(C) The keeper is male on the return journey.**

This question has Eldrick, Bowie, and Demonde on the front bench on the outbound journey. If Eldrick is not the pilot on the outbound journey, our diagram shows that Amani must be the pilot. The outbound journey is "Combination 1" and the return journey is "Combination 2":

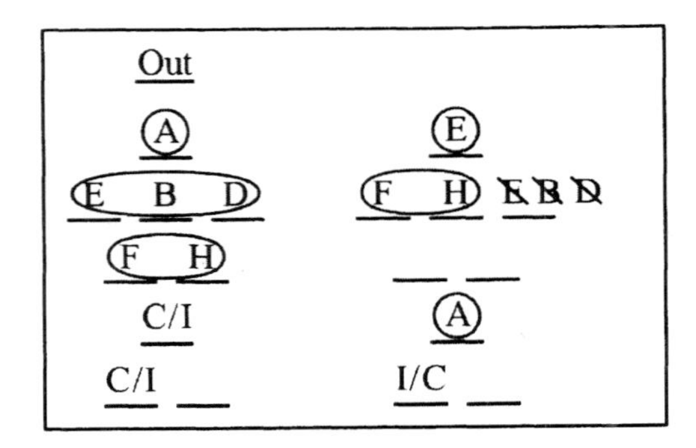

The answer choices to this question all focus on the line keeper. On the outbound journey, the line keeper must be Clifton or Irie since they cannot be on the boat together. However, that means the keeper could be male or female on the outbound journey. Therefore, answer choices (A), (B), and (E) must be eliminated. We do know that Amani, a male, is the line keeper on the return journey, and therefore (D) must be eliminated and answer choice (C) is correct.

5. **(C) All the females sit on the benches on the return journey.**

If the parasail has only females on the outbound journey Irie and Galilea must be in the parasail. The parasail is involved in the rule that either Irie or Clifton have to be seated in the parasail – Irie is a female and Clifton is a male. Galilea must be in the parasail because Frida and Haven only sit in the benches due to their pairing.

Further, if Irie is the parasail we know that Clifton must be in the parasail in the return journey. This leaves only Bowie and Demonde as the other possible entities that can be in the parasail on the return journey.

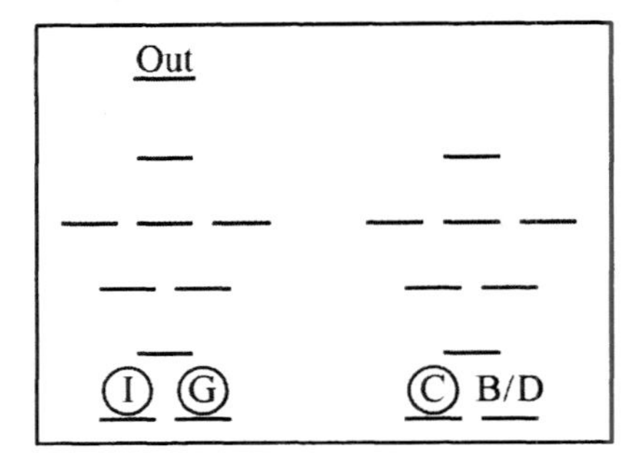

Answer choice (A) is incorrect because there are only males in the parasail on the return journey.

Answer choice (B) is incorrect because Frida and Haven always sit on benches, per the master diagram.

Answer choice (D) is incorrect because Amani, a male, never sits on the benches, per the master diagram.

Answer choice (E) is incorrect because Amani, a male, must always either be the pilot or keeper.

6. **(E) Eldrick sits in a position behind Clifton**

The master diagram and previous work can be used to easily answer this question. The only person that can sit in a position in front of Eldrick is Amani. Eldrick is either in the front bench right behind Amani or is the pilot.

Answer choice (A) is incorrect per “Combination 2” of the master diagram.

Answer choice (B) is incorrect per the outbound journey in question 1. When Amani is the line keeper any entity can be seated in front of him.

Answer choice (C) is incorrect. When Irie is in the parasail, she must be behind Clifton, as they cannot be in the same row.

Answer choice (D) is incorrect per “Combination 1” of the master diagram.

Singing Competition

Relative Ordering GD: Order Entities **Difficulty = Hard**

In this ordering game, you are required to order singers (entities) based solely on relation rules, making this a relative ordering game. This relative ordering game also has a frame with timeslots for spots, lending to the following master diagram

Master Diagram

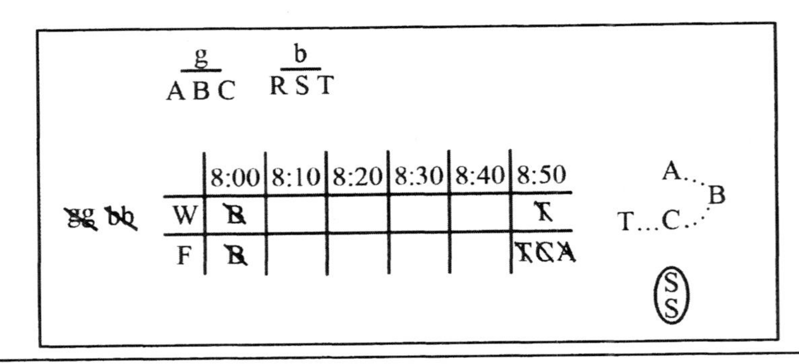

For purposes of brevity the master diagram will not be redrawn when answering the questions.

1. **(A) Rick sings at 8:40 on Friday.**

Since Tom is singing at 8:10 on Wednesday, we know that the two other boys have to sing at 8:30 and 8:50 on Wednesday pursuant to the alternating boy-girl Wednesday rule. This means girls will be singing at 8:00, 8:20, and 8:40 on Wednesday.

Is Tom involved in any other rules? Yes, pursuant to the chain set up to the right of the diagram, Tom has to sing before Claire, and Claire has to sing before Beth. Accordingly, Claire and Beth must take up the last girl spots on Wednesday, which are 8:20 and 8:40 respectively. This leaves Alice with the 8:00 timeslot on Wednesday.

1)		8:00	8:10	8:20	8:30	8:40	8:50
	W	A	T	C		B	
	F						

Answer (B) is incorrect because if Sam sings at 8:40 on Friday he must sing at 8:40 on Wednesday. However, the Wednesday 8:40 timeslot is already taken by Beth.

Answer (C) is incorrect because, per the diagram, Alice sings at 8:00 on Wednesday.

Answer (D) is incorrect because Alice has to sing before Beth, while Beth has to sing after Tom and Claire. Alice can sing no later than 8:00 on Wednesday.

Answer (E) is incorrect because this would break the alternating boy-girl Wednesday rule.

2. **(D) 7**

Since the alternating boy-girl Wednesday rule prohibits starting Wednesday with consecutive girls, if Rick is the first boy to sing in the competition he has to sing at either 8:00 or 8:10 on Wednesday. Accordingly, there are two configurations that need to be tested:.

		8:00	8:10	8:20	8:30	8:40	8:50
2)	W	R	A	T	C	S	B
	F					S	
	W		R		T		C B (crossed out)
	F						

With Rick singing at 8:00 on Wednesday, the remaining boys have to sing at 8:20 and 8:40. The girls will sing at 8:10, 8:30, and 8:50. Tom is the key entity here. Per the diagram, Tom has to sing before both Claire and Beth. This means at least two of the girl timeslots have to be after Tom's timeslot. To make this happen, Tom must sing at 8:20 on Wednesday (since Rick has the earliest boy timeslot).

Putting the rest of the entities into place yields Claire after Tom at 8:30 and Beth in the 8:50 Wednesday timeslots. The 8:10 timeslot is left for Alice and the 8:40 timeslot is left for Sam. Accordingly, all of Wednesday's timeslots are known. On Friday, only Sam's timeslot is known, since he has the same timeslot on both days. Per the diagram, exactly seven timeslots can be determined with this configuration (where Rick sings at 8:00)

In the second configuration Rick sings at 8:10. However, there is a problem that should be immediately apparent based our previous work. Tom needs two timeslots after him for Claire and Beth on Wednesday. The alternating boy-girl Wednesday rule makes this impossible. If Rick sings at 8:10, the earliest Tom could sing is 8:30. Tom singing at 8:30 on Wednesday leaves only one slot for a girl after him (8:40), and therefore Beth would not have a timeslot. Accordingly, the second configuration is not possible.

3. **(B) Tom sings before Rick on Friday.**

If the maximum time is to pass between Alice's two performances, ideally she would be the first singer on Wednesday and last singer on Friday. However, since Beth has to sing after Alice both days, Alice cannot sing last on Friday. The latest Alice can sing on Friday is 8:40.

With Alice in place at 8:00 on Wednesday and 8:40 on Friday the rest of the diagram can be completed. Per the alternating boy-girl Wednesday rule, girls have to be in the 8:20 and 8:40 timeslots on Wednesday. Further, since Claire sings before Beth, it can be concluded that Claire must sing at 8:20 and Beth must sing at 8:40 on Wednesday.

Since Tom has to sing before Claire, he must sing at 8:10 on Wednesday. This leaves Sam and Rick for the remaining Wednesday spots – 8:30 and 8:50. Are these entities involved in any other rules? Yes, Sam sings in the same timeslot on both nights. Since the 8:50 timeslot on Friday is already taken by Beth, Sam cannot sing at 8:50 on either day. Thus, Sam sings at 8:30 on both nights and Rick must sing at 8:50 on Wednesday.

		8:00	8:10	8:20	8:30	8:40	8:50
3)	W	A	T	C	S	B	R
	F				S	A	B

Answer choices (A), (C), (D), and (E) all must be true. Only choice (B) provides an answer that could be false. The 8:00, 8:10, and 8:20 Friday timeslots are open for Tom, Claire, and Rick. However, the order in which Tom, Claire, and Rick sing on Friday cannot be determined (except that Tom has to sing before Claire).

4. **(D) 5**

Per the master diagram, Beth has to sing after Alice, Claire, and Tom. On Wednesday, Beth can only be in slots 8:40 or 8:50 if the alternating boy-girl Wednesday rule is obeyed and Claire and Alice are ahead of her.

On Friday, there is no restriction that prevents girls from singing consecutively. Thus, she can sing in any slots after Alice, Claire, and Tom sing. This would be the 8:30, 8:40, and 8:50 timeslots on Friday. In total, Beth can sing in five timeslots.

5. **(D) The latest Sam can sing on Wednesday is 8:30.**

There are four possible configurations in which Alice sings immediately after Sam on Friday. Sam and Alice only cannot sing on 8:40 and 8:50 on Friday, as this would not leave a timeslot for Beth (who is involved in the rule that she must always be after Alice). Using the four working configurations, the answer choices can be tested.

	8:00	8:10	8:20	8:30	8:40	8:50
W	S	A	T	C	R	B
F	S	A				
W		S				
F		S	A			
W	T	C	S	A	R	B
F			S	A		
W				S		
F				S	A	
W					~~S~~	
F					~~S~~	~~A~~

Answer (A) is incorrect because Beth always has to sing after Alice.

Answer (B) is incorrect because Alice can sing at 8:10 on Wednesday without breaking any rules or limitations.

Answer (C) is incorrect because Tom can sing at 8:00 on Wednesday without breaking any rules or limitations.

Answer (D) is correct because, as explained above, the latest Sam can sing on a Friday is 8:30. Accordingly, the latest Sam can sing on Wednesday is also 8:30 since he has the same timeslot on both days.

Answer (E) is incorrect because Claire can sing at 8:10 on Wednesday without breaking any rules or limitations.

6. **(D) Wednesday: Tom Claire Rick Alice Sam Beth**

Friday: Tom Alice Claire Beth Sam Rick

Answer (A) is incorrect because two boys – Tom and Sam – are in consecutive timeslots on Wednesday.

Answer (B) is incorrect because Beth is not the last girl on Friday.

Answer (C) is incorrect because Sam sings in a different timeslot on Wednesday and Friday.

Answer (E) is incorrect because Claire sings before Tom on Friday.

Heroes vs. Villains

Matching GD: Match Entities **Difficulty = Very Hard**

This is a complicated matching game with three sets of entities – the heroes, the villains, and the weapons. The game is easiest to diagram if you fix the heroes as the spots, the villains as the entities, and then use subscripts for the weapons:

Master Diagram

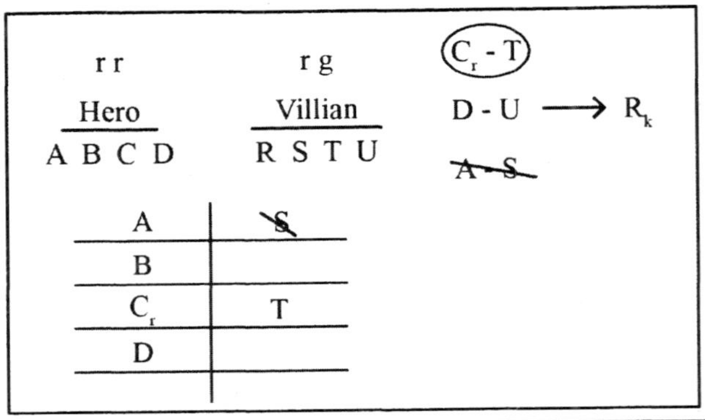

r r — Hero: A B C D

r g — Villian: R S T U

(C_r - T)

D - U ⟶ R_k

~~A - S~~

Hero	Villain
A	~~S~~
B	
C_r	T
D	

1. **(D) Adam and Sam have handguns.**

This question has Bill fighting Ulf with a rifle. If Bill is fighting Ulf, and Carl always fights Tom, Adam must fight Rick (because he cannot fight Sam). This means David and Sam are left to fight each other, and the following diagram results:

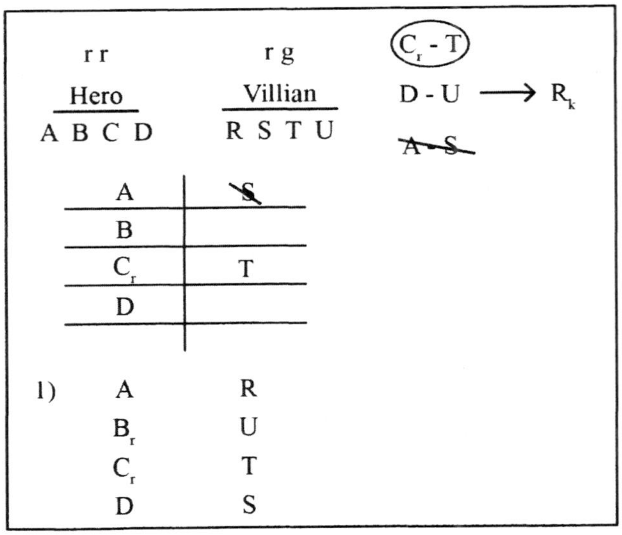

r r — Hero: A B C D

r g — Villian: R S T U

(C_r - T)

D - U ⟶ R_k

~~A - S~~

Hero	Villain
A	~~S~~
B	
C_r	T
D	

1)	
A	R
B_r	U
C_r	T
D	S

Answer choice (A) is wrong because Adam and Rick are fighting each other and, therefore, cannot have the same weapon.

Answer choice (B) is incorrect because Ulf cannot have a rifle, since he is fighting Bill, who already has a rifle.

Answer choice (C) is incorrect because David and Sam are fighting each other and, therefore, cannot have the same weapon.

Answer choice (E) is incorrect because if Rick and Sam have knives, Ulf or Tom would have to be the villain with the rifle. However, both the heroes that Ulf and Tom face (Bill and Carl) already have rifles.

2. **(B) Ulf has a handgun.**

The villains who fight Adam and Bill have handguns in this question. Adam is involved in the rule that he cannot fight Sam. Bill is not directly involved in any rule. If the villains who fight Adam and Bill have handguns, then there are two handguns, a grenade, and a rifle, in total, with the villains. The villains do not have any knives. Since the villains do not have any knives, the conditional rule that results in Rick having a knife cannot be true. Accordingly, David cannot fight Ulf:

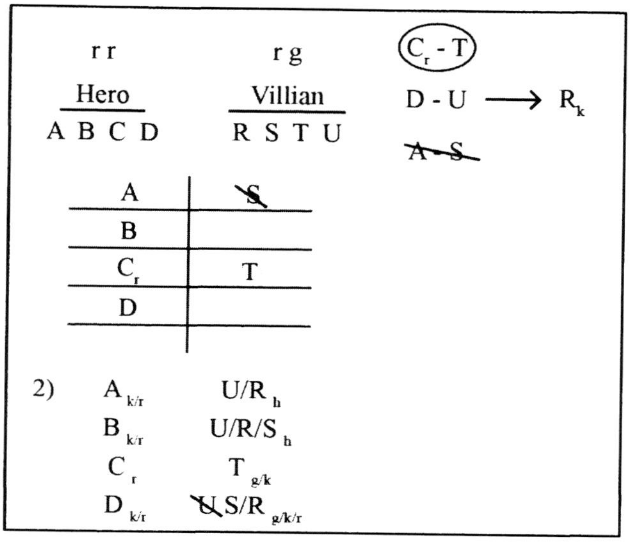

rr   rg   (C_r - T)

Hero   Villian   D - U $\longrightarrow$ R_k

A B C D   R S T U   ~~A-S~~

A	~~S~~
B	
C_r	T
D	

2)	$A_{k/r}$	U/R_h
	$B_{k/r}$	$U/R/S_h$
	C_r	$T_{g/k}$
	$D_{k/r}$	~~U~~ $S/R_{g/k/r}$

This still leaves a diagram showing multiple hero-villain combinations and fighter-weapon combinations. Only answer choice (B), which states Ulf must have a handgun, must be true since Ulf has to fight Adam or Bill. The villains who oppose Adam and Bill have handguns per the limitation set by Question #2. Further (although not drawn in to the diagram) Tom must have the grenade since Carl has a rifle. Thus, whoever is fighting David must have s rifle; and David cannot have a rifle.

3. **(A) 3**

If all the knives are with the heroes, there are several possible weapon pairings for the heroes: Adam could have a knife or a rifle, Bill could have a knife or a rifle, and David could have a knife or a rifle. For the purposes of the question, it is not significant which of the heroes have the knives and which (other than Carl) has the rifle.

The caveat is that, since no knives are left for the villains, the conditional statement that results in Rick having a knife cannot be true. Thus, David cannot fight Ulf. The resulting diagram shows that there are three possible combinations that fit within these limitations:

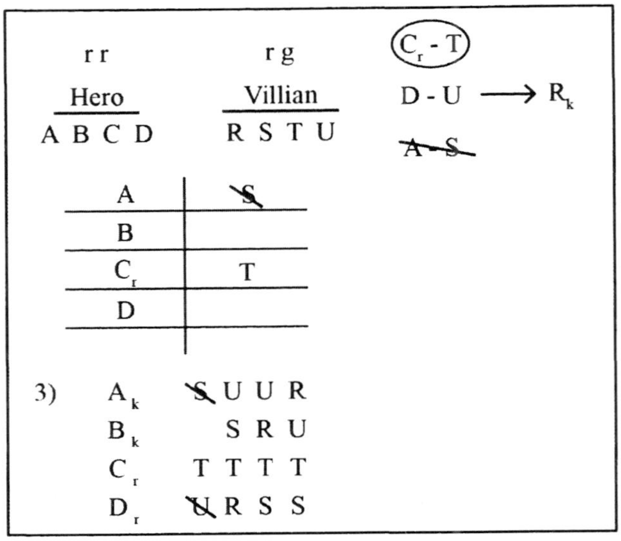

rr — rg — (C_r - T)

Hero: A B C D

Villian: R S T U

D - U ⟶ R_k

~~A - S~~

A	~~S~~
B	
C_r	T
D	

3)

A_k	~~S~~	U	U	R
B_k		S	R	U
C_r	T	T	T	T
D_r	~~U~~	R	S	S

4. **(C) Ulf fights David.**

If Adam has a knife, his opponent cannot have a knife. Adam is involved in the rule that he cannot fight Sam. Thus, Adam's opponent, as always, is either Rick or Ulf. Are Rick or Ulf involved in any other rules? Yes, they are involved in the conditional rule that if David fights Ulf then Rick has a knife. Rick having a knife, however, is a problem if he fights Adam since Adam's opponent cannot have a knife (in this question).

If the conditional statement is true and David fights Ulf, then Rick would have to fight Adam with a knife. Accordingly, David cannot fight Ulf.

rr — Hero: A B C D

rg — Villian: R S T U

(C_r - T)

D - U → R_k

~~A - S~~

Hero	Villian
A	~~S~~
B	
C_r	T
D	

4)

Hero		Villian
A_k	~~S~~	U_h
B_r		R_h
C_r	T	T_g
D_k	~~U~~	S_r

5. **(E) Adam has a knife.**

This question has Bill fighting Sam. Sam is involved in the rule that he cannot fight Adam. Bill is not involved in any rules. Answer choices (A) through (D) allow for two fighter pairings because Adam can only fight Rick or Ulf, and David could fight Ulf or Rick without breaking any rules. Answer choice (E) results in only one possible combination of pairings for the fighers: Adam-Ulf, Bill-Sam, Carl-Tom, and David-Rick.

If Adam has a knife his opponent cannot have one. If David fights Ulf, then in this situation, Rick would have to fight Adam with a knife. Thus, David cannot fight Ulf if Adam has a knife.

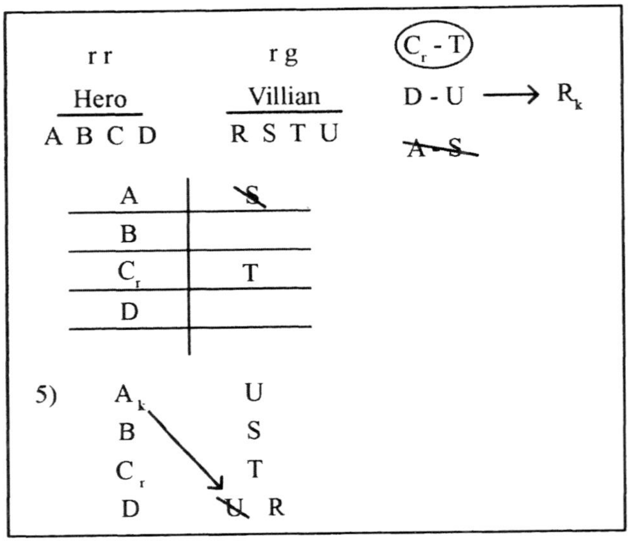

rr — Hero: A B C D
rg — Villian: R S T U

(C_r - T)
D - U → R_k
~~A - S~~

A	~~S~~
B	
C_r	T
D	

5)
A_k — U
B — S
C_r — T
D — ~~U~~ R

Grand Jury Indictments

Matching GD: Match Entities **Difficulty = Hard**

This game requires the matching of executives (entities) to charges (spots). It is fairly straightforward but for one rule. The rule that "The Chief Executive Officer and Chief Financial Officer are indicted for Stock Manipulation and Accounting Fraud, respectively, either together or not at all," is oddly phrased and tricky to decipher. This type of tricky phrasing will be encountered from time to time on the LSAT, and so you must take the rule word by word and diagram accordingly.

The subject rule first states that CEO and CFO are indicted on Stock Manipulation and Accounting Fraud charges "respectively." This means that the CEO is indicted on Stock Manipulation charges and the CFO is indicted on Accounting Fraud charges. The rule then goes on to state "either together or not at all." What does this mean? This means that if you see the CEO with a Stock Manipulation charge you will see the CFO with an Accounting Fraud charge and vice versa. The result is a double arrow as demonstrated in the master diagram below:

Master Diagram

CEO CFO COO GC

Account	Insider	Stock
CFO ⟵	⟶	CEO
COO ⟵	GC	

~~CEO GC~~

1. **(E) The Chief Executive Officer is indicted for Insider Trading.**

Both the GC and the COO are in indicted for Accounting Fraud in this question. Accounting Fraud is involved in two rules in this game. One rule requires the CEO to be indicted for Stock Manipulation if the CFO is indicted on Accounting Fraud.

Since the CFO has not been indicted on Accounting Fraud charges, we know that the CEO cannot be indicted for Stock Manipulation. Further, the CEO cannot be indicted for Accounting Fraud because the maximum number of executives (2) have been indicted for Accounting Fraud. Accordingly, the CEO must be indicted for Insider Trading.

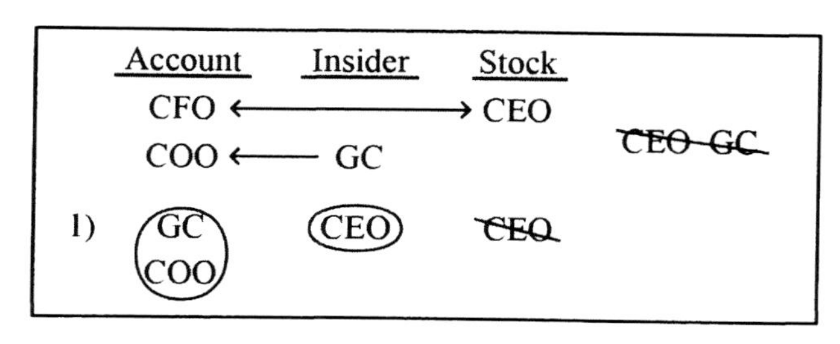

2. **(B) The Chief Financial Officer is indicted on exactly two charges.**

Both the COO and the GC are indicted for Stock Manipulation in this question. Stock Manipulation is in the rule that requires the CFO/CEO to be indicted for Accounting Fraud and Stock Manipulation charges together or not at all. Since both of the Stock Manipulation charges are already taken by the COO and GC, the CEO cannot be indicted on Stock charges and therefore, the CFO cannot be charged for Accounting Fraud. This means that the CFO must be charged with Insider Trading only.

	Account	Insider	Stock	
	CFO ←	———	→ CEO	~~CEO GC~~
	COO ←	— GC		
2)	~~CFO~~	(CEO)	(COO GC)	

Based on the preceding diagram, answer choice (B) is clearly false.

3. **(B) The Chief Operating Officer is indicted on Accounting Fraud.**

If the CFO is indicted on all three charges, we know that CEO must be indicted for Stock Manipulation. This leaves one Insider Trading charge and one Accounting Fraud charge for the COO and the GC.

	Account	Insider	Stock
	CFO ←	———	→ CEO
	COO ←	— GC	
3)	(CFO)	(CFO)	(CFO)
	COO/GC	GC/COO	(CEO)

Answer choices (A) and (E) are incorrect because the CEO is only indicted on Stock Manipulation.

Answer choices (C) and (D) are incorrect because the CEO and CFO are the only executives indicted for Stock Manipulation.

4. **(D) The Chief Executive Office and the Chief Operating Officer**

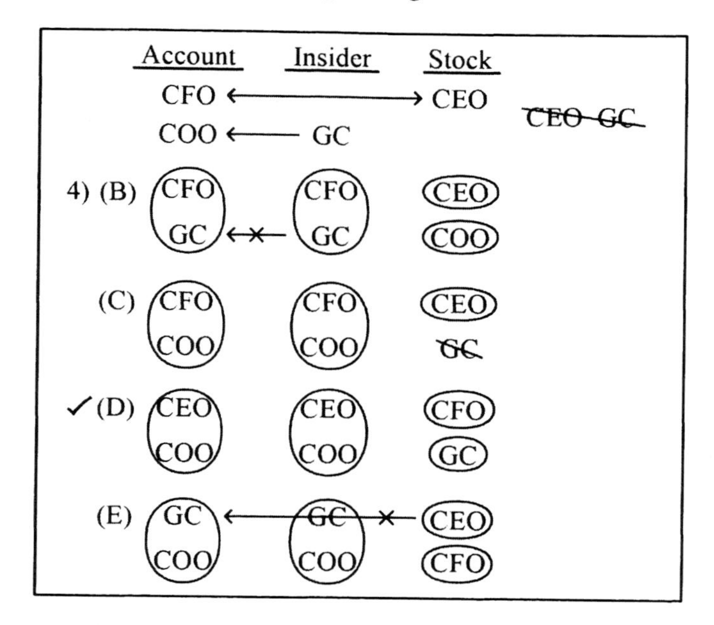

Answer choice (A) is incorrect because the CEO and GC can never be indicted on the same charge.

Answer choice (B) is incorrect because if the GC is indicted for Insider Trading the COO must be indicted for Accounting Fraud.

Answer choice (C) is incorrect because this would result in the CEO and the GC being indicted on the same charge, which is not allowed.

Answer choice (E) is incorrect because CEO would have to be indicted for Stock Manipulation, but the CFO is not indicted of Accounting Fraud. This violates the rule that the CEO/CFO must be indicted of Stock/Accounting together or not at all.

5. **(E) The Chief Operating Officer and Chief Financial Officer are both indicted on exactly one charge each.**

A difficult and not so obvious deduction can be made that because the CEO and GC cannot be indicted on the same charge, the COO and CFO cannot both be indicted on just one charge. If you were not able to deduce this, you can test out all of the answer choices.

Answer choice (A) was true in Question #4.

Answer choices, (B), (C), and (D) can all be true per the following diagram:

	Account	Insider	Stock	
	CFO ←	→	CEO	~~CEO GC~~
	COO ←	GC		
5) (B)	GC	GC	CFO	
	COO	CEO	COO	
(C)	COO	COO	COO	
	GC	CEO	CFO	
(D)	COO	CFO	COO	
	GC	CEO	CFO	

There is no possible arrangement that would allow the COO and CFO to both be indicted on only one charge each. Thus, answer choice (E) is correct.

6. **(C) General Counsel and Chief Executive Officer are both indicted for Accounting Fraud and Stock Manipulation.**

This question removes the rule that the GC and CEO cannot be indicted on the same charge, but leaves all other rules intact. As with the previous two questions, there is no easy way around this question and so we must test each answer choice.

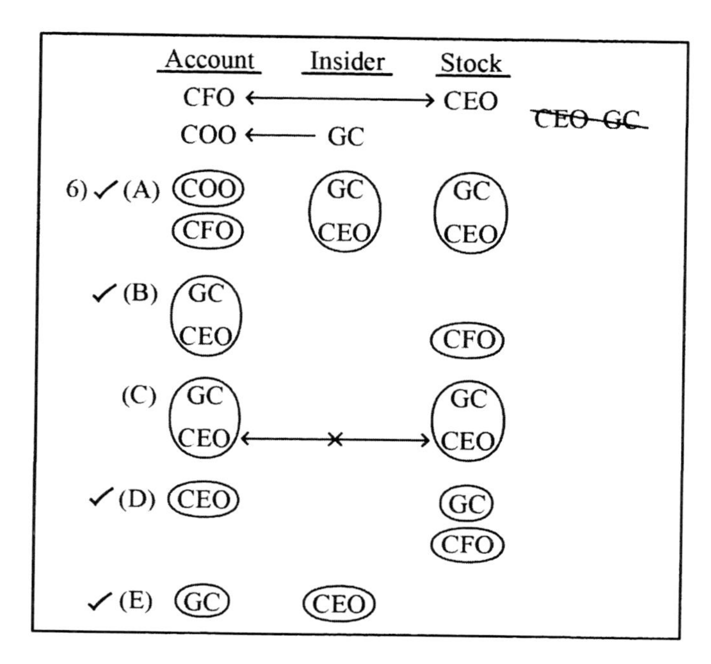

Answer choices (A), (B), (D), and (E) do not violate any rules once the restriction that the CEO and GC cannot be indicted on the same charge is removed. Answer choice (C), however, violates the rule that if the CEO is charged with Stock Manipulation, the CFO must be charged with Accounting Fraud.

Zoo Budget Cuts

Selection GD: Match Entities **Difficulty = Hard**

In this selection game, various animals (entities) must be selected from a larger group of animals. To draw the diagram, start by drawing the rules as provided:

Cats ≤ 4
i → tt
e → 2 rep
sc
ee → s

Several of the rules share entities and therefore can be combined to form new rules. Specifically, the following rules can be combined: the rule that if two elephants are selected then the snake must be selected; the rule that that requires exactly two reptiles to be selected when an elephant is selected; and the rule that the snake and the crocodile cannot be selected together. This results in the deduction that if two elephants are selected then the snake, iguana, and two tigers must be selected:

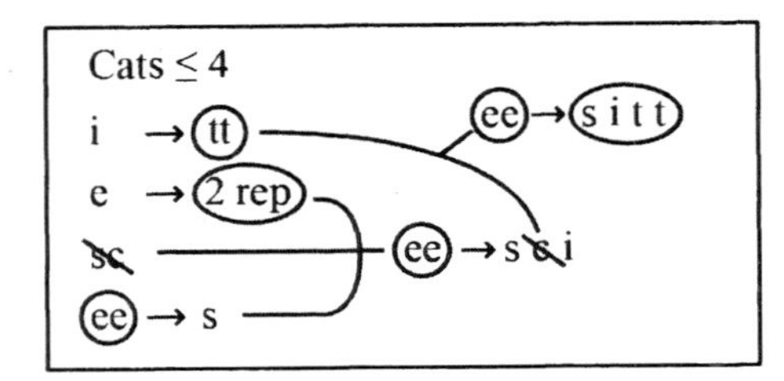

Ultimately, the master diagram should reflect the major deduction that *when two elephants are selected, we know all of the animals that are selected*:

Master Diagram

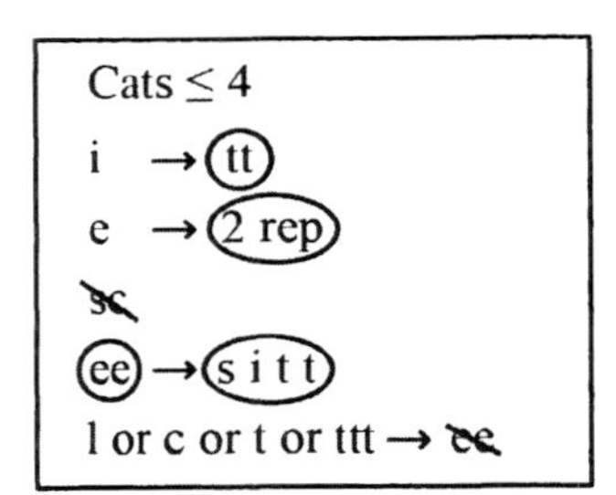

1. **(B) Two tigers, an elephant, two crocodiles, a lion**

Answer choice (A) is incorrect, because the snake cannot be chosen with a crocodile.

Answer choice (C) is incorrect, because when an elephant is chosen two reptiles must be chosen.

Answer choice (D) is incorrect because more than four big cats are chosen.

Answer choice (E) is incorrect because the snake must be chosen if two elephants are chosen.

2. **(D) One lion is chosen.**

The tigers are involved in two rules – that there can be at most four cats, and that if the iguana is chosen exactly two cats are chosen. Thus, if all three tigers are selected then the iguana cannot be selected and only one lion can be selected. Further, we know there cannot be two elephants based on our master diagram. What is left is only two possible configurations:

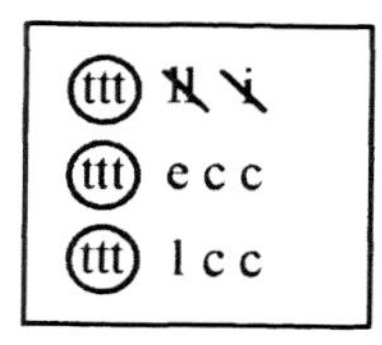

Answer choice (D) could be true, but it is the only answer choice that does not have to be true.

3. **(D) If only one crocodile is chosen exactly two tigers are chosen.**

The rule that involves reptiles is not helpful in this question and, therefore, the various configurations must be tested.

2 rep	others	
c c	t t t l	
c c	t l l e	
c i	t t l l	> answer (D)
c i	t t e l	
s i	t t l l	
s i	t t e e	

Based on the above-diagram answer choice (D) is the only one that must be true.

4. **(A) 1**

The questions states that only one crocodile and one lion are selected. Crocodiles are involved in the rule that states they cannot be chosen with the snake. Lions are involved in the rule that there cannot be more than 4 big cats chosen. Thus, we know for sure that the snake is not selected. Further, two elephants cannot be selected if a lion or crocodile is selected.

With very few deductions that can be made, we must resort to trial and error on this question. Let's start by going through the rules one at a time.

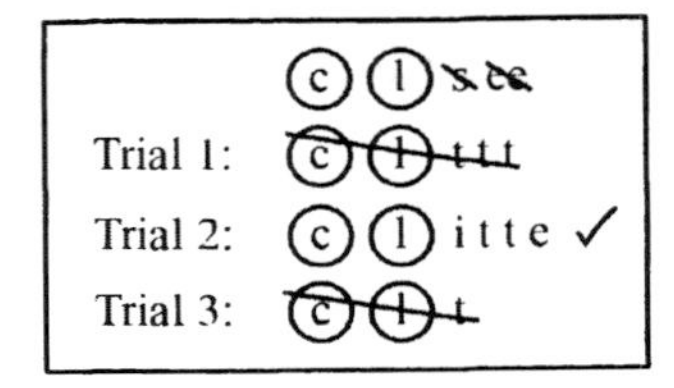

Trial 1: If four big cats are chosen, then three of them are tigers. A crocodile, lion, and three big cats results in five animals selected with only one spot left. An elephant cannot be chosen because two reptiles were not selected. An iguana cannot be selected because there are more than two tigers selected. Thus, there is no other animal that can be chosen for the sixth spot. Accordingly, this configuration does not work.

Trial 2: If an iguana is chosen, then two tigers must be chosen. Five out of six animals are selected. The sixth animal could be an elephant. This works.

Trial 3: Next, let's see if just one tiger can be selected. If exactly one tiger is selected, then the iguana cannot be selected, and the quota for six animals cannot be filled.

There are no other configurations to be tried. Only the configuration from Trial 2 worked. Thus, answer choice (A) is the correct answer.

5. **(D) 7**

This question removes the limitation that snakes and crocodiles cannot be chosen together and is replaced by the rule is that both lions must be selected. Lions are also involved in the rule that there can be no more than 4 big cats. Unfortunately, there are no deductions that will further limit this question, and we must resort to trial and error.

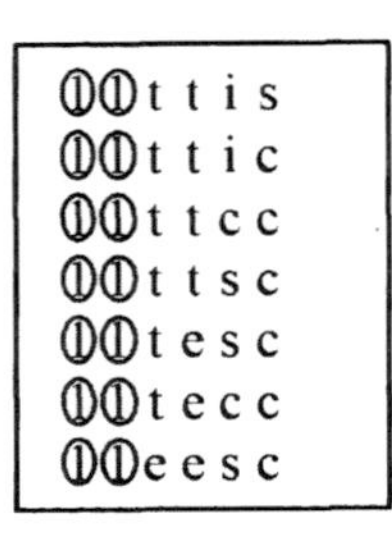

The preceding diagram shows the seven possible ways to arrange the animals when both lions are chosen and the new rule used. Getting rid of the rule that the snake cannot be with crocodiles opens many new possibilities. The best way to approach this is the same way we did in Questions #4 – start with a rule and try the different configurations. Ultimately, seven different configurations are possible.

International Ambassadors

Geometric Positioning GD: Position Entities **Difficulty = Hard**

This is a geometric positioning question where eight entities are sitting around a table, evenly spaced, resulting in an octagon or a circle frame. Accordingly you may either draw an octagon as your frame or a circle with eight evenly spaced marks denoting the respective spots.

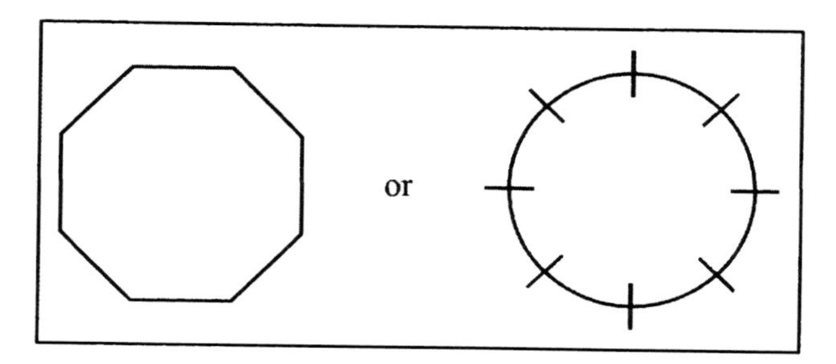

The rules can be drawn as follows:

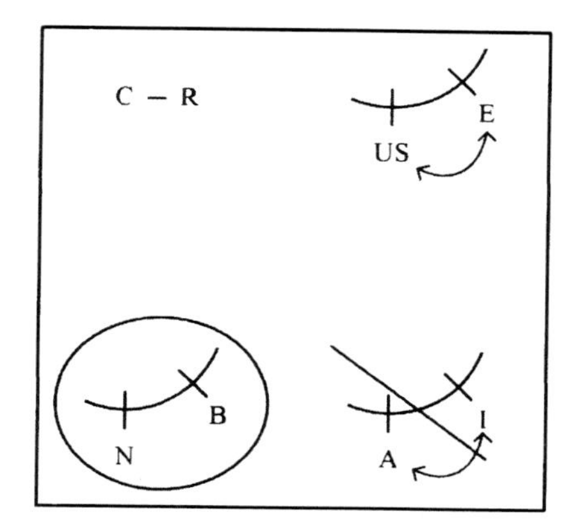

There are no *significant* deductions that can be made in this game. However, keep in mind that because this frame is an evenly spaced circle, even if you drew your rules slightly differently you could still come to the right answer as long as you did so consistently.

For purposes of brevity the master diagram will not be redrawn when answering the questions.

1. **(D) The ambassador from India sits next to the ambassador from England.**

This question has the ambassador from China sitting immediately to the right of the ambassador from the United States. China and the United States are both involved in one rule each. If the ambassador from China sits to the right of the ambassador from the United States, the rules dictate that the ambassador from England must sit to the left of the United States. The ambassador from Russia always sits across from the ambassador from China. The rule that requires the ambassador from Brazil to sit to the right of the ambassador from Nigeria then leaves several spots blocked for Nigeria and Brazil. The following diagram results:

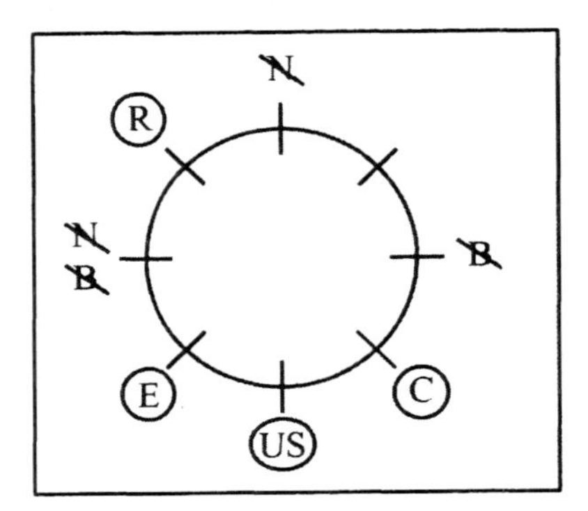

Based on this diagram, only answer choice (D) does not violate any rules.

2. **(A) The ambassador from Brazil sits across from the ambassador from Australia.**

This question has the ambassador from India sitting exactly two seats to the left of the ambassador from Australia. Always start with what the question gives us. In this case, draw India two spots to the left of Australia. Next, determine where the other entities cannot go based on the location of India and Australia. Since China and Russia have to always sit across from each other, they cannot be in the spots across from India and Australia, thus we can cross China and Russia out of these spots.

Further, we know Brazil cannot sit to the immediate right of Australia or India, per the rule that Brazil sits to the right of Nigeria. Thus, we can cross Brazil out of the spots to the right of Australia and India. For the same reason, we can cross out Nigeria from the spots immediately to the left of India and Australia.

We also know that the ambassadors from the United States and England must sit next to each other. Accordingly, they cannot sit in between India and Australia, and we can cross US and England off from that spot. This allows us to deduce that only Russia or China can be in the spot between India and Australia, and also that the spot across from the spot between India and Australia must be either Russia or China.

Lastly, since the spot across from the spot in between India and Australia is occupied by either Russia or China, we can finally deduce that Brazil cannot sit across from Australia, and Nigeria cannot sit across from India. All of the above is incorporated into the following diagram:

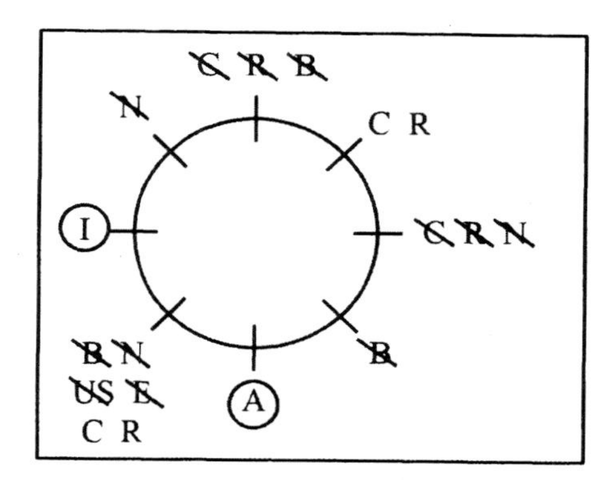

In a "must be false question," all of the incorrect answer choices are ones that could be true. Thus, based on the diagram, all of the answer choices could be true except (A) because the ambassador from Brazil cannot sit across from the ambassador from Australia.

3. **(B) The ambassador from Russia sits next to the ambassador from Brazil.**

This question has the ambassador from Nigeria sitting directly across from the ambassador from India. Nigeria and India are involved in two different rules. Nigeria must sit to the left of Brazil. India cannot sit next to Australia. Neither Russia nor China can sit to the right of India because the seat across from the seat to the right of India is already occupied by Brazil. Further, neither England nor the U.S. can sit between India and Brazil because this would not leave any spots for Russia and China to sit across from each other. Accordingly, the following diagram results:

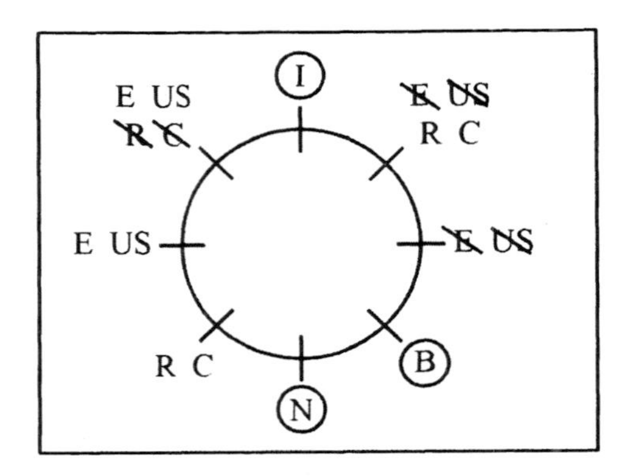

Based on this diagram, all of the answer choices could be true except answer choice (B).

4. **(A) The ambassador from India sits next to the ambassador from Russia or China.**

This question has the ambassador from Australia sitting next to the ambassador from China. Australia and China are involved in two separate rules. We start the diagram by drawing in the rules that Russia is across from China and India cannot be next to Australia. We can then draw in where Nigeria and Brazil cannot go, and the resulting diagram is as follows:

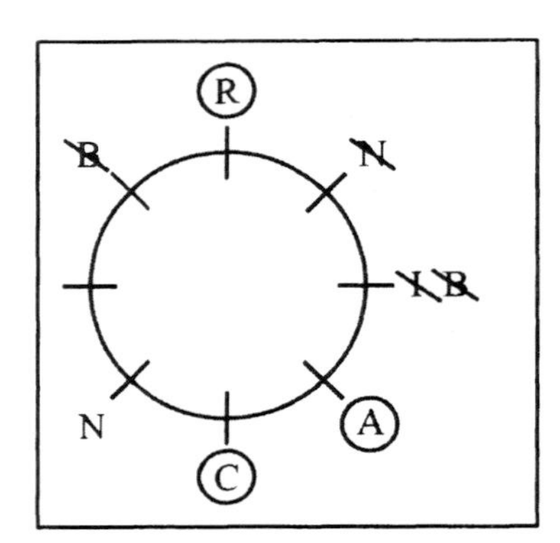

Answer choice (A) is the only one that MUST be true.

5. **(D) 5**

This question removes to the condition that Australia and India are not next to each other. The key to this question is recognizing that that the China/Russia pairing must sit next to the Australia/India pairing, otherwise the Nigeria/Brazil and U.S./England pairings could not sit next to each other. After diagramming this it should be easy to see that the ambassador from England can sit next to all of the other ambassadors besides the ones from Nigeria and Brazil. Thus, the answer is (D) 5.

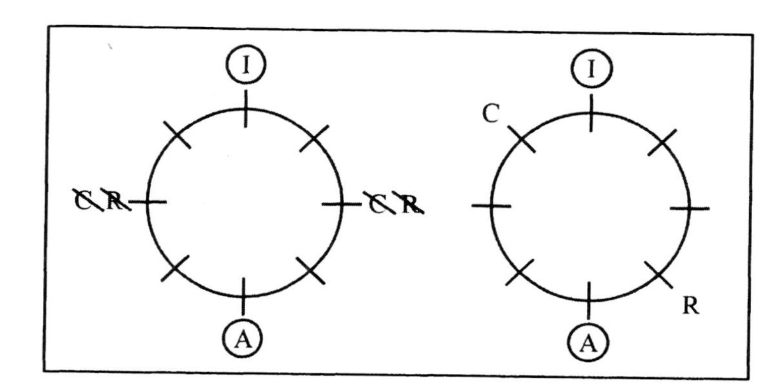

Cheerleading Pyramid

Positioning & Selection GD: Position and Match Entities **Difficulty = Moderate**

This game requires you to place cheerleaders (entities) into positions in a pyramid (spots). There are two possible frames, one for a 4-3-2-1 pyramid and one for a 3-2-1 pyramid.

3 - 2 - 1	4 - 3 - 2 - 1
___	___
___ ___	___ ___
___ ___ ___	___ ___ ___
	___ ___ ___ ___

When the 4-3-2-1 pyramid is used all ten entities are placed. When the 3-2-1 pyramid is used, you must select six entities out of ten for placement. Since there is no ordering/temporal element to this game, the rules are very easy to draw:

(BC) ~~AEHK~~
(DK) J/I = bottom corners

Note that ~~AEHK~~ usually means that A, E, H, and K cannot all be placed together in a row. For sake of brevity, and because the rules in this game are not complex, it is acceptable to draw this as ~~AEHK~~.

Remember to scan the rules and combine rules with common entities. In this case, Kiki is involved in two rules lending to the deduction that Deedee also cannot be with in the same row as Ally, Ellen, and Helen. However, for this particular game we do not need to separately draw this deduction, as it captured well in the diagram already. Another major deduction in this game is that A, E, H, and K must each be on separate rows. Your final diagram, therefore, should like this:

Master Diagram

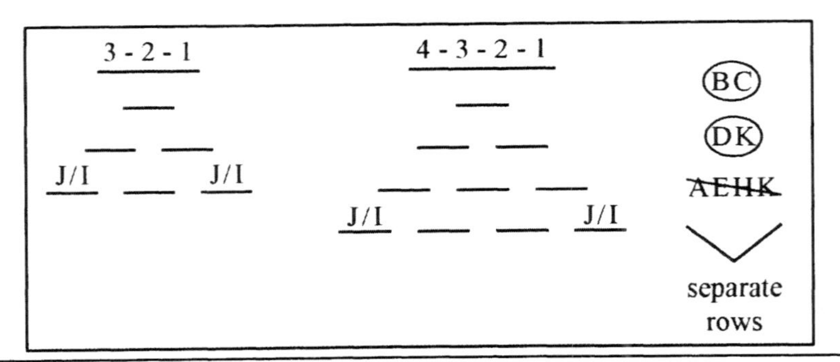

1. **(B) Helen and Ellen are in the pyramid but Fifi is not.**

This question has Ally at the top of a 3-2-1- pyramid. Per the master diagram, the pyramid will have three spots occupied and three spots open:

A
___ ___
I ___ J

Ally is only involved in one rule, but it is of no significance to this question, as she is at the top of the pyramid. The key to this question is that all of the wrong answers involve the BC or DK pairings. Answer choice (B) is the only one that does not.

If Bibi, Cici, Deedee, or Kiki are present they must be in the middle row so as to satisfy their pairing rules. Accordingly, the answer choices can be easily diagramed as follows:

(A)
A
K D
I H J

(B)
A
H ~~BCDK~~
I E J

(C)
A
K D
I E J

(D)
A
K D
I E/H J

(E)
A
B C
I E/F/H J

Answer choice (B) must be false because if Helen and Ellen are both in this pyramid, they must be on different rows. Further, Fifi must be in the second row, as neither the BC nor DK pairings would fit.

2. **(C) Fifi is in the second row.**

This question has Kiki next to Igor in the pyramid. Kiki is involved in two rules, one where she has to always be next to Deedee and the other where she cannot be next to Ally, Ellen, or Helen. The fact that Kiki has to be next to Deedee, per the rule, and Igor, per the question, means she must be in the bottom row of a 4-3-2-1 pyramid. Since all ten entities must be used in this question, Ally, Ellen, and Helen have to take one spot each on the top three rows, this results in the following diagram:

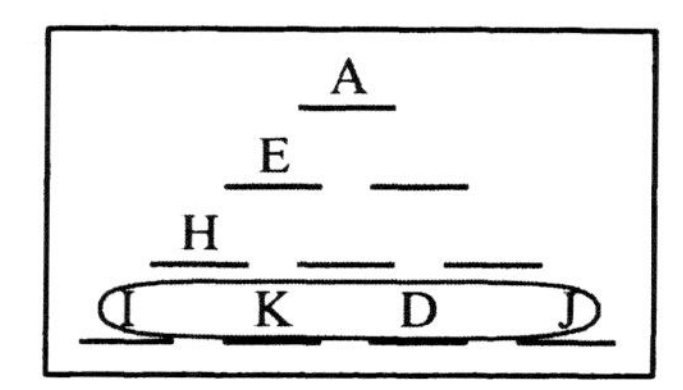

Further, we know that Bibi and Cici must always be next to each other; therefore they must go in the row with 3 spots. This leaves Fifi in the row with 2 spots:

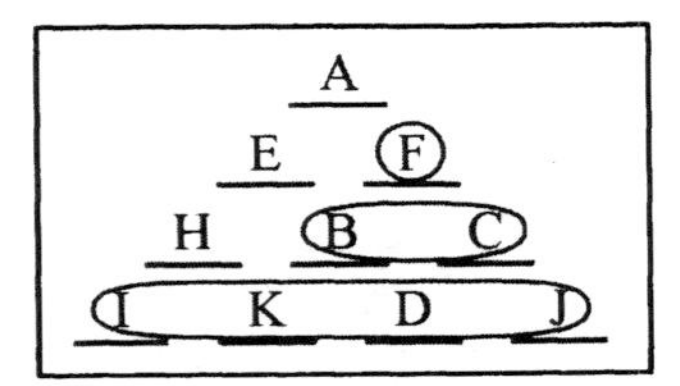

Accordingly, answer choice (C) must be true. Note also that answer choices (B) and (E) should have also been eliminated immediately as both Ally and Helen have the same properties and, therefore, the answers are interchangeable (They both have to be correct or both have to be wrong. But since there is only one correct answer, they both have to be wrong).

3. **(D) Helen and Fifi are in the third row of a 4-3-2-1 pyramid.**

This question has Helen and Fifi in the row below Ellen. Helen and Ellen are both involved in the rule that they cannot be in the same row as Ally or Kiki. This question requires some trial and error, by placing Ellen in each row and Helen and Fifi in the row below:

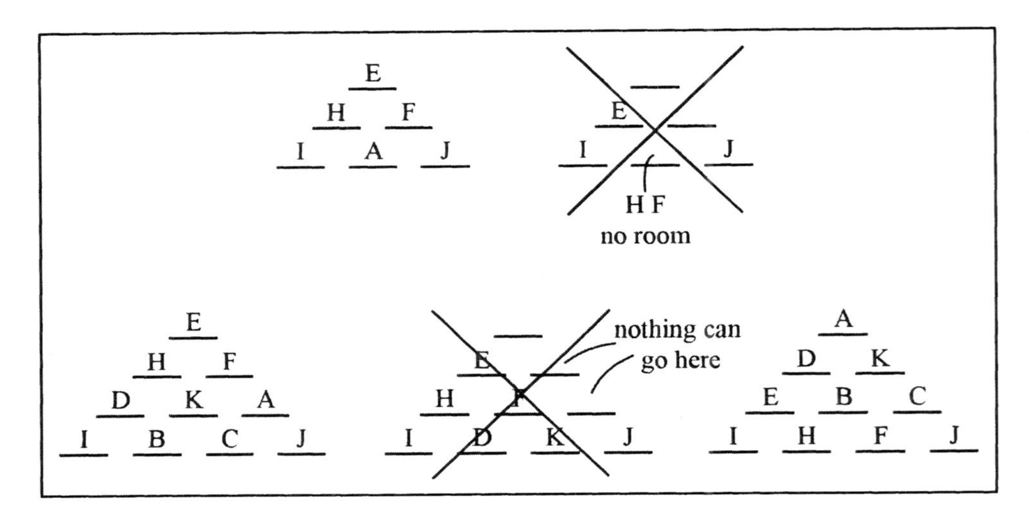

Per the preceding diagram, E cannot be in the second row and HF can never be in the third row. Accordingly, answer choice (D) is correct.

4. **(E) Kiki and Deedee are on the second row.**

This question can be answered based solely on our previous work.

Answer choice (A) is incorrect because Fifi can never be at the top as only Ally, Ellen, or Helen can be at the top per the rule that Ally, Ellen, Helen, and Kiki cannot be in the same row.

Answer choice (B) is incorrect as Kiki can never be at the top because Kiki must always be in the same row as Deedee.

Answer choice (C) is incorrect as because if Bibi and Cici are in the bottom row with Igor and Jon, there will be no room for Ally, Ellen, Helen, or Kiki, who must all be on different rows.

Answer choice (D) is incorrect because if Kiki and Deedee are in the third row, then the BC pairing would have to go in the bottom row or the second row. However, if the BC pairing is in the bottom row or second row, then the rule that says Ally, Ellen, Helen, and Kiki cannot be in the same row will inevitably be violated:

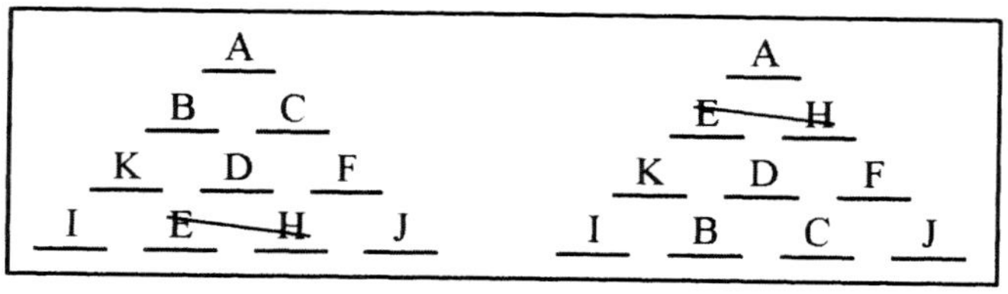

A
B C
K D F
I ~~E~~ ~~H~~ J

A
~~E~~ ~~H~~
K D F
I B C J

5. **(E) If neither Kiki nor Bibi is in the pyramid, then Helen must be in the pyramid.**

This question can be answered using prior work.

Answer choice (A) is incorrect per the following diagram.

A
E F
I H J

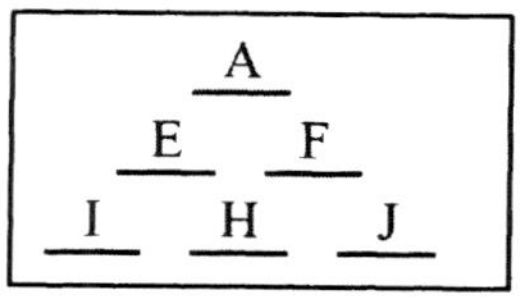

Answer choices (B) and (C) are also incorrect per the preceding diagram.

Answer choice (D) is incorrect per the following diagram.

H
B C
I E J

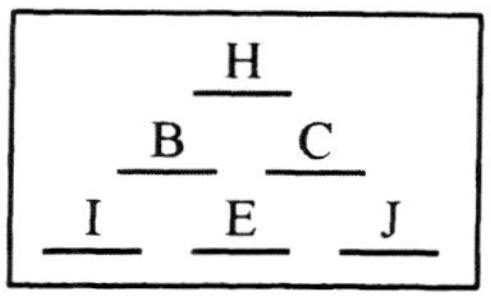

Answer choice (E) is correct because if neither Kiki nor Bibi are in the pyramid, then neither are their respective pairings, Deedee and Cici. This means that the remaining six entities must be in the pyramid.

6. **(C) Helen**

Since Ally, Ellen, Helen, and Kiki have to be on different rows, we can deduce that one of these four must be on the top row. Thus, answer choices (A), (B), and (D) can be eliminated immediately. Answer choice (E) is also incorrect because Kiki cannot be at the top of the pyramid since she is involved in the rule that she must be next to Deedee, and there is only one spot at the top of the pyramid.

7. **(D) Bibi, Cici, Fifi, Jon**

This question requires you to ignore the rule Igor and Jon must be on outside corners of the bottom row. With all other rules still applying, the bottom row would still need to have either Ally, Ellen, Helen, or Kiki. Answer choice (D) is the only that does not contain any of these entities.

Real Celebutantes

Selection GD: Match Entities **Difficulty = Moderate**

In this selection game you must select five out of ten potential cast members (entities). The diagram in this game is not easy to visualize, and is best drawn by connecting the entities that are feuding with lines to create what looks like a tree diagram:

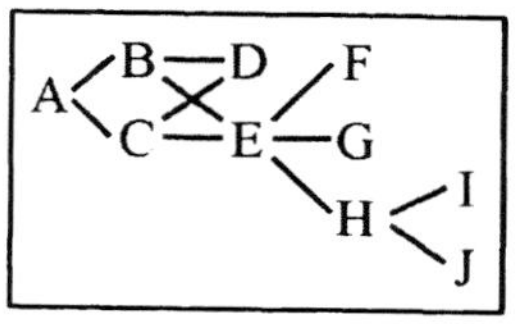

A B D F C E G H I J

Next, since the alliances are formed by entities with common enemies, in the preceding diagram of the feuds, every other block of entities forms an alliance group. Three alliance groups result: DEA, BCFGH, and EIJ. Accordingly, your final diagram should look like this:

Master Diagram

Feuds	Alliances
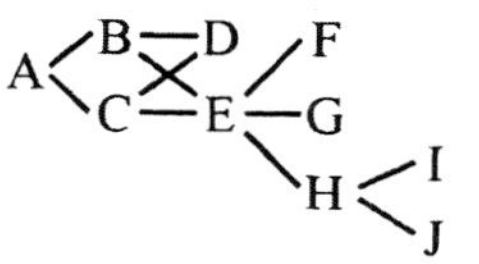 A B D F C E G H I J	D E A B C F G H E I J

When answering questions, you must check to ensure that each entity that is selected has at least one feud and at least one alliance. This can be accomplished by drawing a chart for each question, as will be displayed below.

1. **(C) Calista and Hester**

	D	E	F
Feud	B/C	✓	✓
Alliance	✓	✓	G/H

The question states that Dionne, Ella, and Fallon are selected. Dionne already has an alliance in Ella, but she does not have a feud, as she feuds with either Brittany or Calista. Ella's feud is already satisfied by Fallon, and her alliance is satisfied by Dionne. Fallon's feud is satisfied by Ella, and her alliance can be fulfilled by Brittany, Calista, Gigi, or Hester.

2. **(B) Brittany and Ella**

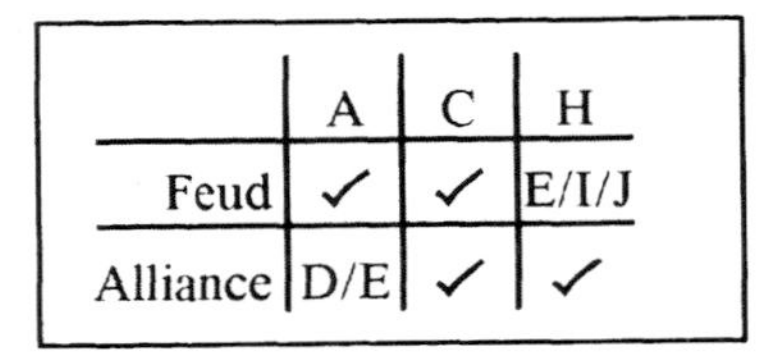

	A	C	H
Feud	✓	✓	E/I/J
Alliance	D/E	✓	✓

The question has Alexia, Calista, and Hester selected for the cast. Alexia's feud is satisfied by Calista, but she needs Dionne or Ella for an alliance. Calista's feud is satisfied by Alexia and her alliance is satisfied by Hester. Hester needs Ella, Iman, or Juliet for her feud, but her alliance is already satisfied by the presence of Calista.

Thus, the chart shows that either Dionne *with* Iman or Juliet must be selected or Ella with any other entity (that does not violate any rules) must be selected. Answer choice (B) is correct because Ella can be selected with Brittany, as Brittany's alliance is satisfied by Calista and her feud is satisfied by Alexia.

Answer choices (A) and (C) are incorrect because either Iman or Juliet must be selected if Dionne is selected.

Answer choice (D) and (E) are incorrect because neither Dionne nor Ella are selected.

3. **(E) The selection is unacceptable because Fallon has no feud.**

This question requires you to determine if each of the selected entities are matched with a feud and an alliance. Starting with Fallon first, it is clear that she does not have a feud. Accordingly, (E) is the correct answer.

4. **(D) Ella**

This question has Alexia selected for the cast. Based on the master diagram if Alexia is chosen, then either Brittany or Calista are chosen as Alexia's feud. However, Brittany and Calista are equivalent entities in this game – they have the same exact properties (the same feuds and alliances). Thus, in a must be true question, equivalents can be immediately eliminated as wrong answers. Accordingly, answer choices (A) and (B) are incorrect. Answer choice (E) can also be eliminated, as Hester has no direct relationship with Alexia, per the master diagram.

That leaves only Dionne and Ella. Either one clearly *can be* selected per the master diagram; but just one must be selected. First let's try a combination where Dionne is not selected:

Alexia is selected and feuds with Brittany and has an alliance with Ella (instead of Dionne). Ella can also feud with Brittany and has an alliance with Alexia. Fallon can feud with Ella and have an alliance with Brittany. Finally, Gigi can feud with Ella and have an alliance with Brittany.

This works. That means Dionne does not *have* to be selected, and answer choice (C) is incorrect and (D) is the only answer left. To be sure that (D) is the correct answer, let's give it a try:

Alexia is selected and feuds with Brittany/Calista and has an alliance with Dionne (instead of Ella). Dionne can also feud with Brittany/Calista and has an alliance with Alexia. But now who can be selected?

Fallon or Gigi cannot be selected because they have no feuds if Ella is not selected. If Hester is selected she will have an alliance with Brittany/Calista but would need Iman or Juliet for a feud. However, neither Iman nor Juliet can be selected because they only have alliances with each other or Ella. Thus, Ella is absolutely necessary if Alexia is chosen.

5. **(A) Brittany, Calista, Ella, Fallon, Gigi, or Hester**

This is a giveaway question if you properly diagramed the alliances, as displayed in the master diagram. Recall that three groups of alliances are formed in this game: DEA, BCFGH, and EIJ.

6. **(D) Brittany, Calista, Ella, Hester, Iman**

As always with acceptability questions, you will need to eliminate the answer choices that violate the rules to find the correct answer.

Answer choice (A) is incorrect because if Alexia is chosen, either Brittany or Calista must be chosen.

Answer choice (B) is incorrect because if Juliet is chosen Hester must be chosen.

Answer choice (C) is incorrect because neither Dionne nor Iman have alliances.

Answer choice (E) is incorrect because there is no feud for Dionne.

7. **(A) Alexia, Brittany, Calista, Dionne, Ella**

The new rule for this question is that each cast member must now have two feuds. Based on the master diagram, Fallon, Gigi, Iman, and Juliet can have only one feud each. Thus, any answer choices that have those entities must be wrong. Using this deduction we can immediately eliminate answer choices (C), (D), and (E).

Answer choice (B) is incorrect because Hester doesn't have any feuds.

The Puma

Matching (Distribution) GD: Match Entities **Difficulty = Hard**

This is a visually complex matching game, wherein you are tasked with matching the contestants (entities) with three date types (spots). Further you must also distribute four to five Carnations (properties) to the entities.

There are a couple of major deductions that can be made to greatly simplify this game. First and foremost, since Yardley, Rico, and Zane cannot be on the same date, each date type must have one of those three contestants. Yardley can only be in the Solo and Group Dates and always gets a Carnation. This can be drawn directly into the diagram. Further, Rico can only get Carnations if he is on the Group Date or the 2-on-1 Date, and gets eliminated if he is on the Solo Date. This can also be drawn directly into the diagram. Zane can go on any date and may or may not get a Carnation on any date he goes on.

Second, because Wes and Xavier must go on the same date, they must be on the Group Date. They cannot go together on the Solo Date because only one contestant is on the Solo Date. Further, they cannot go together on the 2-on-1 Date, as either Rico or Zane must also be on that date.

The rules and above deductions can be diagrammed several different ways. We have chosen a table format, which allows you to keep track of those entities receiving a Carnation and those who are eliminated.

Master Diagram

R T U V W X Y Z

	① Solo	Group	2-1
✓	Y	③ Y/R	① R
X	R	②	①
	Z	(WX) Z TUV	Z TUV

~~YRZ~~

V ↔ U

For purposes of brevity the master diagram will not be redrawn when answering the questions.

1. **(A) Upton gets a Carnation on the 2-on-1 Date.**

If Zane is eliminated on the Group date, then Yardley cannot be on the Group date, as Yardley is involved in the rule that says Zane and Yardley do not go on the same dates. This would mean Yardley must be on the Solo date, as Yardley can only be on the Group or Solo dates. As a result, Rico would have to go on the 2-on-1 Date, where he gets a Carnation.

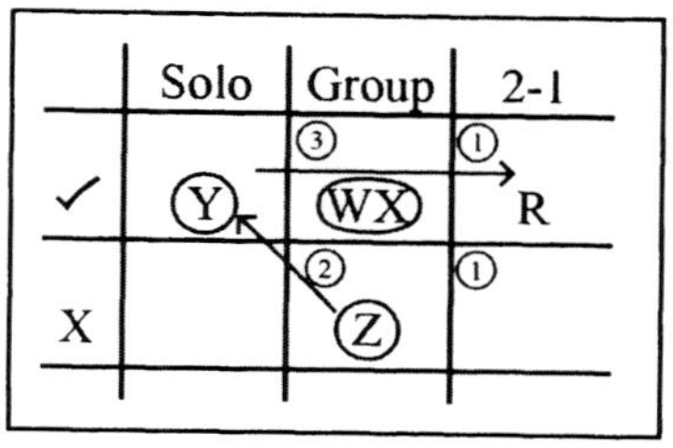

	Solo	Group	2-1
✓	Y	③ WX	① R
X		② Z	①

Accordingly, Upton cannot get a Carnation on the 2-on-1 Date as Rico takes the only Carnation on that date.

2. **(E) Zane**

The question has Tate and Vaughn in the Group Date. This means we know four of the five entities on the Group Date – Tate, Vaughn, Wes, and Xavier (as Wes and Xavier always go on the Group Date). Further, either Yardley, Rico, or Zane must be on the Group Date. This leaves Upton for the 2-on-1 Date.

Furthermore, since Vaughn is eliminated, Upton must receive a Carnation. This means Upton *must* receive a Carnation on the 2-on-1 date. Looking at our master diagram, we can see that Rico can only be on the 2-on-1 date if he receives a Carnation. That means Zane must be on the 2-on-1 Date and that he gets eliminated.

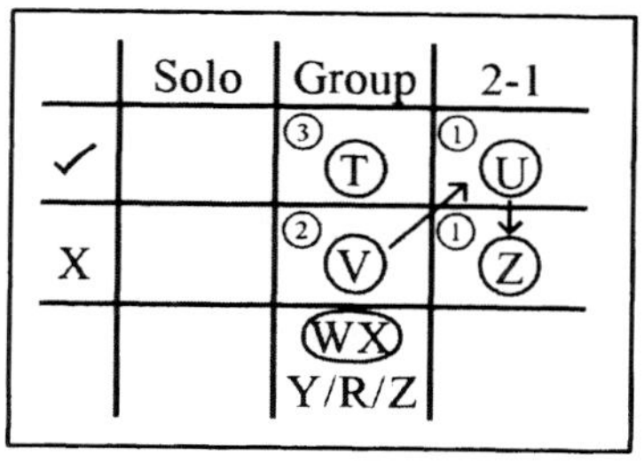

	Solo	Group	2-1
✓		③ T	① U
X		② V	① Z
		WX Y/R/Z	

3. **(B) Rico, Zane**

Per our master diagram, only Yardley, Rico, and Zane can be in the Solo Date. However, Yardley always gets a Carnation.

4. **(C) Yardley, Wes, Xavier, Upton**

Answer choice (A) is incorrect because either Upton or Vaughn must get a Carnation per the rule that one gets a Carnation and one is eliminated.

Answer choice (B) is incorrect because both Upton and Vaughn cannot receive Carnations per the rule that one gets a Carnation and one is eliminated.

Answer choice (D) is incorrect as the maximum number of contestants that gets Carnations is five – three from the Group Date, one from the 2-on-1 Date, and one from the Solo Date.

Answer choice (E) is incorrect because Yardley always gets a Carnation.

5. **(C) Tate goes on the Group Date.**

This question has Rico, Wes and Xavier eliminated. Per the master diagram, if Rico is eliminated he must be in the Solo Date. This means Yardley must be in the Group Date, and gets a Carnation as always. Further, Zane must then be on the 2-on-1 Date, as he cannot be on the same dates as Yardley or Rico. Accordingly, answer choices (A), (B), and (E) can be quickly eliminated.

Further, since the only spots open in the Group Date are for contestants who get Carnations, both Upton and Vaughn cannot go on the Group Date (because one of those two must be eliminated per the rules). That means Tate must go on the Group Date, where he will get a Carnation.

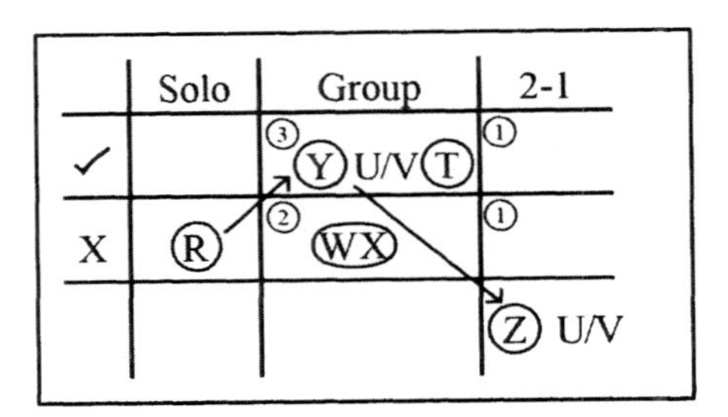

6. **(B) Rico is eliminated on the Solo Date.**

If Tate gets a Carnation on the 2-on-1 Date, then Zane must be eliminated on the 2-on-1 Date, as neither Yardley nor Rico can be eliminated on 2-on-1 Date. Accordingly, answer choice (E) can be eliminated, as Zane cannot be on the Solo Date.

Further, since the 2-on-1 Date has been filled, Upton and Vaughn must be on the Group Date. And since one of them must be eliminated, either Wes or Xavier (but not both) must also be eliminated.

	Solo	Group	2-1
✓		③ U W/X	① (T)
X		② V W/X	① (Z)

Thus answer choices (C) and (D) must be eliminated. Finally, answer choice (A) can be eliminated because Rico can never be eliminated on the Group Date.

7. **(E) If there are four eliminations, then Yardley is on the Group Date.**

The answer choices for this question surround the number of eliminations in the episode. Four eliminations occur when Rico or Zane are eliminated on the Solo Date. Three eliminations occur when Yardley or Zane are given Carnations on the Solo Date. To see what else must be true, let's diagram a configuration with three eliminations and a configuration with four eliminations.

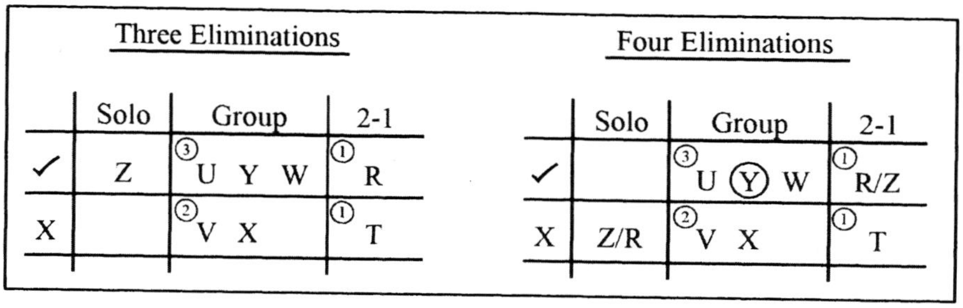

Three Eliminations

	Solo	Group	2-1
✓	Z	③ U Y W	① R
X		② V X	① T

Four Eliminations

	Solo	Group	2-1
✓		③ U Ⓨ W	① R/Z
X	Z/R	② V X	① T

Per the diagram, when there are four eliminations, Yardley must get a Carnation on a Group Date. Answer choice (E) is the only one that must be true.

Speed Dating

Matching GD: Match Entities **Difficulty = Hard**

This a complex matching game which is especially hard to manage because of the number of entities and properties involved. When encountering these types of games, it is often helpful to set up your diagram as a table with the various properties as the rows and columns. The master diagram uses the smoking properties as rows and the political view properties as columns. Women are on the top internal row of each box and men are on the bottom internal row. Since each entity has one smoking property and one political view property, the entities will fall into exactly one of the boxes:

Women: A B C D E F G

Men: T U V W X Y Z

	regular	social	non
liberal	A V		C E X Z
moderate		G T	
conservative	B U W		D F Y

Each box in this table indicates possible date pairings. For example, Anya can date Victor, Bree can date Ulrich or Will, etc. Only Greta and Thomas, who are in the middle of the table as moderate social smokers, do not have any restrictions on whom they can date. Thus, each dater must be paired with someone in his/her box or Greta/Thomas. Establishing this table makes attacking the questions *significantly* easier.

1. **(E) Anya Dates Victor, Bree dates Ulrich, Cerise dates Xavier**

The answer choices to this acceptability question each start with Anya. The table shows that Anya must date either Victor or Tomas. Thus, answer choices (C) and (D) can be eliminated immediately.

Next, the answer choices have various pairings for Bree. Answer choice (B) has Bree paired with Youssef. The table shows that this is not possible. Thus, answer choice (B) is incorrect.

Answer choice (A) is incorrect as Cerise cannot be paired with Youssef, per the master diagram.

2. **(C) Xavier, Zubin, Tomas**

This is a very easy question if you have set up the table as we have in the master diagram. Xavier and Zubin are the same box as Emmy and Tomas can be paired with any girl.

3. **(E) 4**

This question has Greta on a date with Youssef, and asks how to maximize the number of people who do not end up with dates. The way to do this is figure out how to set up Greta and Tomas with dates so that each box has more girls than boys and/or more boys than girls. Doing this will leave potential daters without dates.

If Greta dates Youssef, Danielle and Fiona will not have dates. If Tomas dates Bree, Ulrich and Will won't have dates. Thus, it is plausible that four people could end up with no dates.

4. **(C) Bree, Greta, Tomas, Youssef**

Greta and Tomas always get dates because they have no restrictions. When Greta and Tomas are paired with a liberal, then any one of the liberal boys and/or liberal girls could end up with no dates. When Greta and Tomas are paired with conservatives, Bree and Youssef will still have dates, no matter what.

Answer choice (C) is correct because Bree can always be paired Will if Ulrich is taken by Greta, and vice versa. Likewise, Youssef can always be paired with Danielle if Fiona is taken, and vice versa.

5. **(E) Greta dates a conservative.**

This question has all of the conservative regular smokers getting dates. Bree, Ulrich, and Will are the regular smoking conservatives. If they all get dates, then Bree must date either Ulrich or Will and Greta must date the one Bree does not date. This means that Greta must date a conservative (Will or Ulrich)

Answer choice (A) is incorrect, as Ulrich can date Bree, a conservative.

Answer choice (B) is incorrect, as Will can date Bree, a conservative.

Answer choice (C) is incorrect, as Bree must date a conservative.

Answer choice (D) is incorrect, as Tomas has no restrictions.

Whale Watching

Undefined GD: Match Entities **Difficulty = Very Hard**

This is an example of a game that does not neatly fit into any game types described in this book. While other test takers may panic by this fact, you should recognize that this game can be diagrammed just like all other games using a basic frame with spots (the seats) and entities (the friends).

Master Diagram

G H I JK

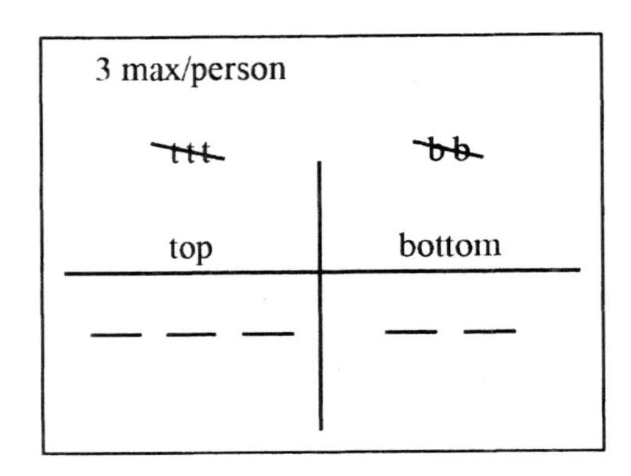

This game is quite difficult because not only are you facing an unfamiliar look and feel to the game, but the questions also require a considerable amount of time and work. Games like this will not appear often on LSATs, but when they do you must be prepared to attack them by diagramming what you know, and carefully testing the answer choices. This game is illustrative of why it is so important to have a streamlined and methodical approach that will allow you to breeze through the easier games, to provide extra time for difficult games.

For purposes of brevity, the rules will not be reproduced in the diagrams below.

1. **(E) Hermes, Jacques, Katriane**

If Gaston, Hermes, and Iphigenie are on top deck on the first trip, then Jacques and Katriane are on the bottom deck for that trip. Since no entities can be on the bottom deck twice in a row, Jacques and Katriane must be on top deck during the second trip. None of the other answer choices have both Jacques and Katriane, so (E) must be the correct answer.

	top	bottom
1)	G H I	J K
2)	J K	

2. **(C) Iphigenie**

The question tells us who is on the top deck of the first and fourth trips, and so the first step is to diagram the information provided:

	top	bottom
1)	G H K	
2)		
3)		
4)	H I	

If Gaston, Hermes, and Katriane are on top deck during the first trip, then Iphigenie and Jacques are on the bottom deck in the first trip. On the second trip, Iphigenie and Jacques must be on top deck because of the rule that states no one is allowed to sit on the bottom deck for two consecutive trips.

	top	bottom
1)	G H K	I J
2)	I J	
3)		
4)	H I	

Per the diagram, Iphigenie is on top deck on the second and fourth trips. This implicates the rule that no one can be in the top deck for three consecutive trips. In order to avoid Iphigenie being in the top deck for three straight trips, she must be in the bottom row for the third trip. Thus, answer choice (C) is correct:

	top	bottom
1)	G H K	I J
2)	I J	
3)		I
4)	H I	

3. **(B) Iphigenie and Katriane are on the top deck in the fifth trip.**

This is a very complicated question that requires some thinking outside the box. As always, begin with what is known. The question states that, on the first trip, Iphigenie and Katriane are on bottom deck. Thus, Gaston, Hermes, and Jacques are on top deck.

	top	bottom
1)	G H J	I K

Because of the rule that no one can be on bottom deck twice in a row, Iphigenie and Katriane will be on top deck during the second trip. It is unknown which of the entities is also on the top deck with Iphigenie and Katriane, but because we will need to keep track of that entity for this question, let's represent him or her with an open circle. Then let's represent the two entities on bottom deck on the second trip with a square and a triangle.

	top	bottom
1)	G H J	I K
2)	I K○	□△

Since the answer choices discuss which entities are on the top deck for the third, fourth, and fifth trip, we will need to continue with the above analysis. Whichever entity is on top deck during the second trip will be implicated by the rule that no one can be on top deck more than twice in a row, sending him to bottom deck on the third trip. The two entities on the bottom during the second trip must be on top deck during the third trip so they don't violate the rule of no consecutive trips on bottom deck. As for Iphigenie and Katriane, one of them will remain on top deck during the third trip and the other will be on the bottom deck. For purposes of differentiation, let's call one of them K/I and the other I/K:

	top	bottom
1)	G H J	I K
2)	I K○	□△
3)	I/K□△	○K/I

Continuing this analysis for five trips results with the following diagram:

	top	bottom
1)	G H J	I K
2)	I K○	□△
3)	I/K□△	○K/I
4)	K/I ○□	I/K△
5)	K/I I/K△	□○

Based on the preceding diagram, answer choice (B) is the only one that MUST be true.

4. **(E) Iphigenie, Jacques, Hermes**

This question has Gaston, Katriane, and Jacques on top deck for the first trip. Katriane is on top deck on the second trip, and Iphigenie is on top deck during the third trip. As always your diagram should be filled with the information provided in the question:

	top	bottom
1)	G K J	H I
2)	K	
3)	I	

The next three trips can then be easily filled in based upon the rules:

	top	bottom
1)	(G K J)	H I
2)	(K) H I	G J
3)	(I) G J	K H

Once the first three trips are determined, we must get creative again to figure out the fourth and fifth trips. Katriane and Hermes were on bottom deck during the third trip, so they must be on top during the fourth trip to avoid violating the rule that no one should be on bottom deck twice in a row. Iphigenie was on top deck on the second and third trips so she must be on bottom during the fourth trip to avoid violating the rule that no one should be on top more than twice in a row.

Katriane must also be on the bottom deck during the fifth trip so as not to violate the rule that no one spends more than three trips on top deck. Iphigenie was on bottom deck during the fourth trip, so she must be on top deck during the fifth trip. Of Gaston and Jacques, one will be on top during the fourth trip and on bottom during the fifth, and the other will be on bottom during the fourth trip and on top during the fifth.

	top	bottom
1)	G K J	H I
2)	K H I	G J
3)	I G J	K H
4)	K H △	O I
5)	H I O	△ K

Based on the above, Hermes, Iphigenie, and either Jacques or Gaston must be on top deck during the fifth trip. Accordingly, answer choice (E) is correct.

5. **(C) Minimum five, Maximum five**

This question involves two rules. The boat will keep making trips until everyone has seen a whale. It takes three times on top deck to see a whale, and no one who has seen a whale can be on top deck afterward. Now take a look back at the past diagrams you have created for other questions. It takes a minimum of four trips for any *two people* to experience three times on top deck (since no entity can be on top deck more than twice in a row). But looking at the past work, you can see that it takes five trips for all five people to be on top deck three times, and thus to have seen a whale. The boat will make exactly five trips to sea, no more and no less. Accordingly, answer choice (C) is correct.

6. **(E) Exactly one of them is on top deck during the fifth trip.**

This is another question that can be answered with prior work. Let's use the most recent diagram (from Question #4) as a starting point.

Of the three people on top deck during the first trip, we can see that one of them will be on top deck during the second trip, and two will be on top during the third trip. This eliminates answer choices (A) and (C). Moreover, two of them will be on top during the fourth trip, eliminating answer choices (B) and (D). However, exactly one of them will be on top deck during the fifth trip, confirming answer choice (E) as the correct answer.

Bathroom Line

Assigned Ordering GD: Order Entities **Difficulty = Very Hard**

This game requires you to place six people (entities) in a bathroom line. The rules are somewhat complicated and the questions are difficult, making this is a very hard game. The frame and rules are diagramed below.

Taco ABC Bar DEF

1 2 3 4 5 6

B...E
D...C
BF
D_F

There are several deductions that can be made in this this game. First, Bhumika, Divya, and Falguni are involved in multiple rules, and so the rules can be combined. The ultimate result is a symbolized version of the rules drawn below.

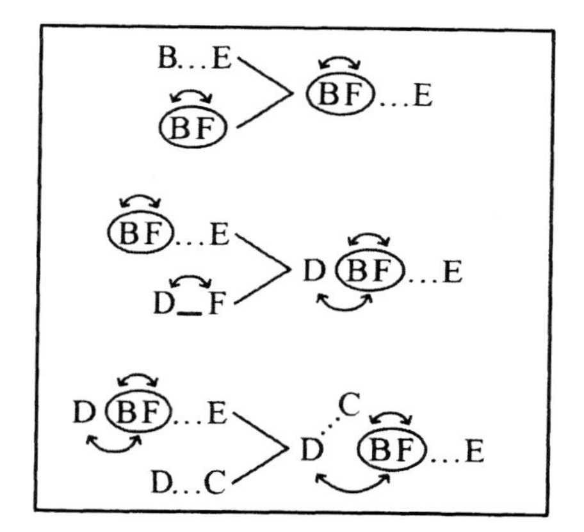

Now, the above depiction of the rules is not clearest or easiest to work with on the test. This is not our goal when diagramming. Although we do want to try to combine as many of the rules as possible, we want to ensure that our diagrams are still easy to read. Thus, in this game, you should decide at what point you want to stop combining the rules.

Ultimately, the rules lend to four main configurations, with Anuradha being a "floater" with no restrictions.

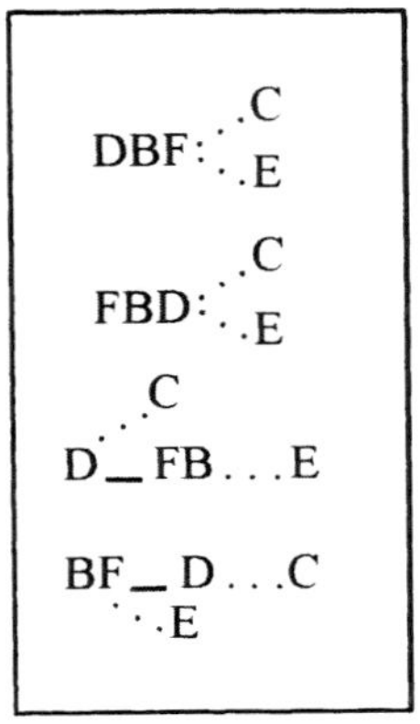

```
      .C
DBF:
      .E

      .C
FBD:
      .E

  .C
D_FB...E

BF_D...C
  ..E
```

Although it is not imperative that you recognize the four main configurations, figuring this out early will help you speed through the questions.

Lastly, remember to draw in the special deductions for the first and last spots of assigned ordering games. In this game, Chaaya and Eashwar cannot be first, and Divya, Bhumika, and Falguni cannot be last. The most helpful master diagram will then look as follows:

Master Diagram

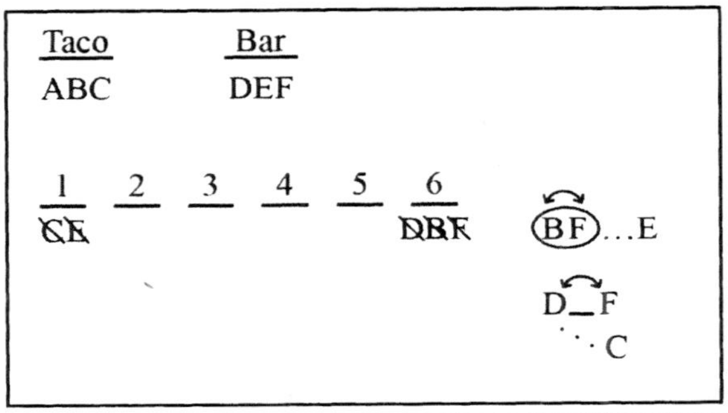

```
Taco        Bar
ABC         DEF

 1   2   3   4   5   6
 _   _   _   _   _   _
CE                  DBF      (BF)...E

                             D_F
                              ..C
```

(In slot 1, C and E are crossed out; in slot 6, D, B and F are crossed out.)

1. **(A) Bhumika is first in line.**

This question has Anuradha third in line. If Anuradha is third, then we know that neither Divya nor Falguni can be in spots 1 or 5 because this would lead to one of those entities being spot 3, which is already occupied by Anuradha. Thus, the Divya/Falguni pairing must go in spots 2 and 4.

The following diagram shows the configuration when Falguni is second, and the configuration when Divya is second. When Falguni is second, no rules are violated, but when Divya is second no entities can go in spot 1.

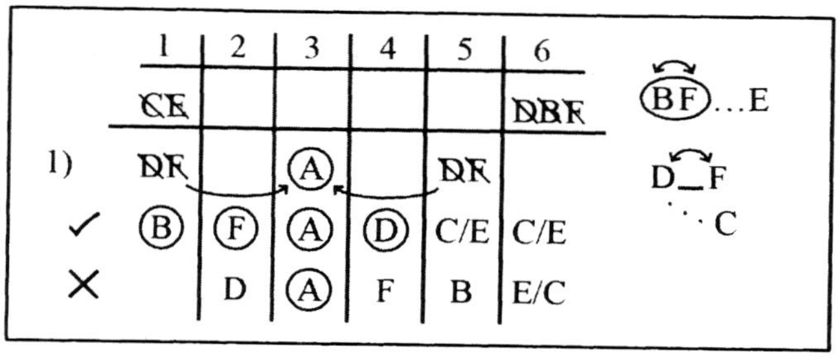

	1	2	3	4	5	6
	~~CE~~					~~DBF~~
1)	~~DF~~		(A)		~~DF~~	
✓	(B)	(F)	(A)	(D)	C/E	C/E
✗		D	(A)	F	B	E/C

(BF)...E
D_F
∴ C

Only answer choice (A) could be true, as Falguni must be in spot 2 and Bhumika must be next to her in spot 1.

2. **(B) Third**

Only Bhumika and Falguni must be in line ahead of Eashwar.

3. **(C) Divya is fifth in line.**

This question requires us to test each choice.

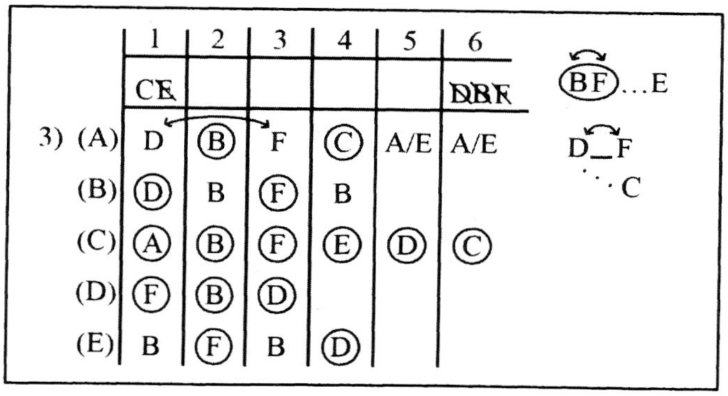

	1	2	3	4	5	6
	C~~E~~					~~DBF~~
3) (A)	D	(B)	F	(C)	A/E	A/E
(B)	(D)	B	(F)	B		
(C)	(A)	(B)	(F)	(E)	(D)	(C)
(D)	(F)	(B)	(D)			
(E)	B	(F)	B	(D)		

(BF)...E
D_F
∴ C

Answer choice (C) is correct because if Divya is fifth, then Chaaya must be last, as Chaaya is always sometime after Divya. Further, since one spot always separates Divya and Falguni, Falguni must be in spot 3. And since Eashwar comes sometime after Bhumika and Falguni, Eashwar must be in spot 4. Bhumika must be in spot 3 because she must be immediately next to Falguni. Spot 1, therefore, is left for Anuradha.

Answer choice (A) is incorrect because if Chaaya is fourth, only Bhumika's spot can be determined with certainty.

Answer choice (B) is incorrect because if Divya is first, only Falguni's spot can be determined with certainty.

Answer choice (D) is incorrect because only the first three spots can be determined with certainty.

Answer choice (E) is incorrect because only Divya's spot can be determined with certainty.

4. **(D) Anuradha is third in line.**

This question has the bar patrons - Divya, Falguni, and Eashwar – first and last in line. Of the bar patrons, only Eashwar can be last. Thus, Divya/Falguni pairing must go in spots 1 and 3.

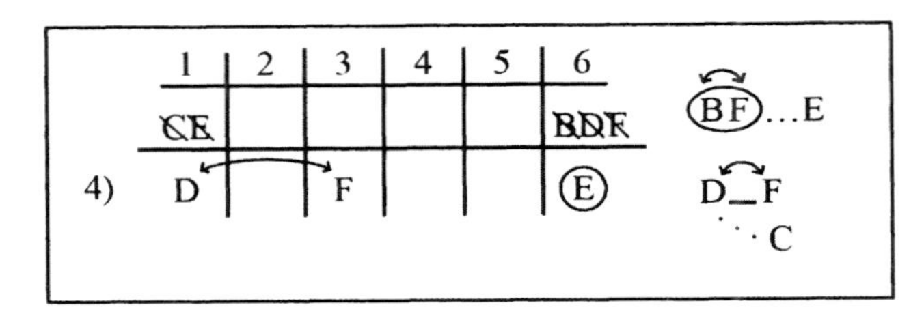

Since spot 3 is now occupied by either Divya or Falguni, any answer choice that has an entity in spot 3 must be false. Accordingly, answer choice (D) is false.

5. **(E) Anuradha is sixth in line.**

This question has the Bar patrons -- Divya, Falguni, and Eashwar – in line consecutively. Since we previously deduced that Falguni must always come before Eashwar, the lineup must have Falguni before Eashwar, and Eashwar before Divya: FED. Further, since Bhumika is always next to Falguni, we can deduce that Bhumika, Falguni, Eashwar, and Divya are together as a group, in that order. This results in two configurations:

	1	2	3	4	5	6
	CE					BDF
5)	B	F	E	D	A/C	A/C
	A	B	F	E	D	C

BF...E

D_F ... C

Only answer choice (E) could be true based on the preceding diagram.

6. **(E) Eashwar is fifth in line.**

This question has the Taco customers - Anuradha, Bhumika, and Chaaya - in spots 2, 4, and 6. That means the Bar patrons - Divya, Eashwar, and Falguni - can only be in spots 1, 3, and 5. Eashwar has to be fifth because Falguni must always come before Eashwar, and Falguni and Divya are always separated by one spot.

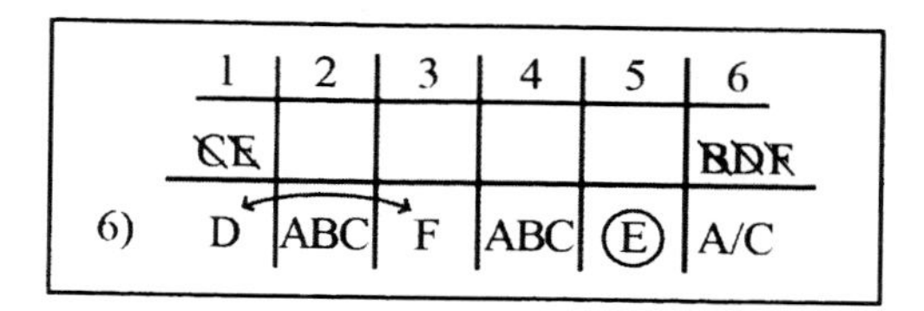

7. **(D) Divya, Anuradha, Chaaya, Falguni, Bhumika, Eashwar**

Answer choice (A) is incorrect because Eashwar is before Bhumika.

Answer choice (B) is incorrect because Chaaya is before Divya.

Answer choice (C) is incorrect because Falguni and Bhumika are not consecutive.

Answer choice (E) is incorrect because Divya and Falguni are not separated by one spot.

Appendices

APPENDIX A – FORMAL LOGIC

DEDUCTIVE REASONING

Deductive Reasoning is the process of using general statements or premises to reach specific conclusions. Here is an example:

All humans are mammals.
I am a human.
Therefore, I am a mammal.

If the rules of deductive logic are followed and the premises are true, then the conclusion must necessarily be true.
Thus, if you properly use deductive logic on the LSAT, and the premises are true (which will be on the LSAT) then your conclusions will also be true. [This is an example of deductive logic!]

The LSAT tests deductive reasoning. Be careful not to confuse deductive reasoning with *inductive reasoning*, which is the process of reasoning from specific examples to reach general conclusions. Inductive arguments can be strong or weak, but deductive arguments can be valid or invalid. The LSAT tests whether you can make valid deductive arguments.

INTERPRETING IF-THEN STATEMENTS

Formal logic, in the realm of the LSAT, is a system of deductive reasoning. It is the primary concept tested in the LSAT Logic Games section. On test day, you will be presented if-then statements, such as "***If*** Alice dances first ***then*** Beth dances fourth." The statement "if A is 1st, then B is always 4th" is a deductive formal logic statement.

If-then statements are conditional statements. If a certain condition is met, a result will occur.

If-then statements can be presented in a variety of ways. Here are several different ways "***If*** Alice dances first ***then*** Beth dances fourth" can be written:

- ***If*** Alice dances first ***then*** Beth dances fourth.
- Alice dances first ***only if*** Beth dances fourth.
- Alice dances first ***if, and only if***, Beth dances fourth.
- Beth ***always*** dances fourth ***if*** Alice dances first.
- ***When*** Alice dances first, Beth dances fourth.
- Alice dances first **only when** Beth dances fourth.
- ***If*** Alice dances fourth, Beth ***must*** dance fourth.

All of the above statements are equivalent and mean the same exact thing. Notice that, although these are all considered "if-then" statements, they do not all contain the words "if" or "then."

All of the above statements can be represented very easily in a diagram using a simple arrow. "***If*** Alice dances first ***then*** Beth dances fourth" is represented symbolically as A → B. At the back end of the arrow is the "if" entity, and the front end of the arrow is the "then" entity. A is the trigger, and B is the result. Thus, no matter how the logical statement is presented, it can be simply diagrammed as:

A → B

CONTRAPOSITIVE

The "contrapositive" is an inverted manner to express a conditional statement. Every conditional statement has a contrapositive. A conditional statement and its contrapositive are logically equivalent, i.e. they mean the same thing.

For example, the logical statement "***If*** Alice dances first ***then*** Beth dances fourth," can also be written in these inverted *contrapositive* forms:

- ***If*** Beth ***does not*** dance fourth, ***then*** Alice ***does not*** dance first.
- ***If*** Beth ***does not*** dance fourth, ***then*** Alice ***cannot*** dance first.
- ***When*** Beth does not dance fourth, Alice ***does not*** dance first.

Proper use of the Contrapositive is a critical skill that will either simplify your game setup or jeopardize the entire diagram. If a logical rule is true, then its contrapositive must also be true.

We can demonstrate this by using a contrapositive example from real life. Consider the following rule:

If the sun is visible in the sky, then it must be daytime.

S→D

In order to get the contrapositive, **negate both statements *and* flip the arrow:**

Not D → not S

If it is not daytime, the sun is not visible in the sky.

But proceed with Caution: DO NOT negate the statements without flipping the arrow and DO NOT flip the arrow without negating the statements!!! If you make an improper negative deduction, the result would be:

~~Not S → not D~~

~~If the sun is not visible, then it cannot be daytime.~~

Our real life example demonstrates the limits of a proper contrapositive. If the sun is visible in the sky, it must be daytime. If it is not daytime, then the sun is not visible in the sky.

However, just because the sun is not visible in the sky does not necessarily mean that it is not daytime. The sun could be obstructed by fog, or covered by clouds and rain. It is entirely possible for it to be daytime but for the sun to not be visible.

Use contrapositives properly and wisely to make deductions!

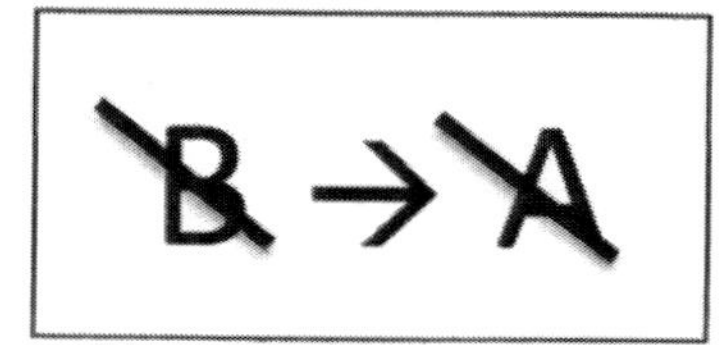

APPENDIX B – SYMBOLOGY

Using uniform symbols to represent logic is key to success in the Logic Games section. There are many different ways in which you can visually represent logic that is provided on the LSAT, and there is no "correct" way to symbolize the logic. We recommend using the symbols below, as they will allow you to maintain consistency throughout.

Symbol	Translation	Usage	Ordering	Non-Ordering
Arrow A → B	If-Then	Represents an if-then or trigger-result relationship. The arrow can go either any direction, with the "if" or trigger on the back end of the arrow, and then "then" or result going on the front end of the arrow.	If Al shoots first, then Bud shoots fourth: 1 2 3 4 5 6 7 A → B	If Al drinks wine, then Bud also drinks wine: Wine Beer H20 A → B
Circled Entity (A)	Always goes in designated spot.	Represents a *must be true* or always true logical statement.	One of the girls finishes third: 1 2 3 4 5 6 7 (g)	Bud drinks wine: Wine Beer H20 (B)
Multiple Circled Entities (AB) Ordering Variation: (AB)	Ordering: Entities are consecutive, *as drawn*. Non-Ordering: Entities always go together.	Ordering: Represents the relationship that two entities must be consecutive, as drawn. If the entities are consecutive but not in a specific order, then a double arrow must be placed above to indicate this. Non-Ordering: Represents that two entities must always go together in the game. If one is a group, then the other must also be in that group. If one is selected the other must be selected, etc.	Al shoots exactly one spot before Bud. (AB) Al and Bud shoot consecutively: (AB)	Al and Bud drink the same beverage. (AB)

Crossed-Out Entities ~~A~~	Cannot go in designated spot.	Represents a *cannot be true* logical statement.	No girl finishes third: 1 2 3 4 5 6 7 ~~G~~	Bud does not drink beer: Wine Beer H20 ~~B~~
Multiple Crossed-Out Entities ~~AB~~ Ordering Variation: ~~AB~~	Ordering: Entities not consecutive, *as drawn*. Non-Ordering: Entities cannot go together.	Ordering: Represents the relationship that two entities cannot be consecutive, as drawn. Non-Ordering: Represents that two entities cannot go together. If one is in a group, then the other cannot be in that same group. If one is selected the other cannot be selected, etc.	Al does not shoot in spot before Bud. ~~AB~~ Al and Bud do not shoot consecutively: ~~AB~~	Al and Bud drink different beverages. ~~AB~~
Three Dots A ... B	One entity comes sometime after another entity.	Represents the relationship of one entity relative to another.	Al shoots sometime before Bud: A ... B	*Not used*
Dashes A __ B A __ __ B A __ __ B	Entities are separated by exact number of spots, *as drawn.*	Represents how many spots are between entities. The number of spots between entities is represented the number of dashes. One dash means one entity separates the	Al shoots exactly two spots before Bud: A __ B Exactly two players shoot in between Al and Bud: A __ __ B	*Not used*